CONTENTS

For more on these titles, visit: www.independence.co.uk

A note on critical evaluation

Because the information reprinted here is from a number of different sources, readers should bear in mind the origin of the text and whether the source is likely to have a particular bias when presenting information (just as they would if undertaking their own research). It is hoped that, as you read about the many aspects of the issues explored in this book, you will critically evaluate the information presented. It is important that you decide whether you are being presented with facts or opinions. Does the writer give a biased or an unbiased report? If an opinion is being expressed, do you agree with the writer?

Tackling Climate Change offers a useful starting point for those who need convenient access to information about the many issues involved. However, it is only a starting point. Following each article is a URL to the relevant organisation's website, which you may wish to visit for further information.

south essex college

FURTHER & HIGHER EDUCATION
SOUTHEND CAMPUS

Tackling Climate Change

ISSUES

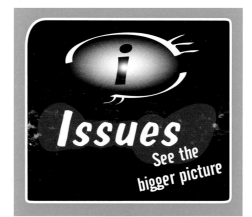

Volume 216

Series Editor

Lisa Firth

Independence

Educational Publishers
Cambridge

First published by Independence

The Studio, High Green

Great Shelford

Cambridge CB22 5EG

England

© Independence 2011

British Library Cataloguing in Publication Data

Tackling climate change. -- (Issues ; v. 216)

1. Global warming. 2. Climatic changes.

I. Series II. Firth, Lisa.

363.7'3874-dc23

ISBN-13: 978 1 86168 596 4

Printed in Great Britain

MWL Print Group Ltd

Climate and climate change: some background science

An extract from The Royal Society's report Climate change: a summary of the science.

The greenhouse effect

The Sun is the primary source of energy for the Earth's climate. Satellite observations show that about 30% of the Sun's energy that reaches the Earth is reflected back to space by clouds, gases and small particles in the atmosphere, and by the Earth's surface. The remainder, about 240 Watts per square metre (W/m^2) when averaged over the planet, is absorbed by the atmosphere and the surface.

To balance the absorption of 240 W/m^2 from the Sun, the Earth's surface and atmosphere must emit the same amount of energy into space; they do so as infrared radiation. On average the surface emits significantly more than 240 W/m^2, but the net effect of absorption and emission of infrared radiation by atmospheric gases and clouds is to reduce the amount reaching space until it approximately balances the incoming energy from the Sun. The surface is thus kept warmer than it otherwise would be because, in addition to the energy it receives from the Sun, it also receives infrared energy emitted by the atmosphere. The warming that results from this infrared energy is known as the greenhouse effect.

Measurements from the surface, research aircraft and satellites, together with laboratory observations and calculations, show that, in addition to clouds, the two gases making the largest contribution to the greenhouse effect are water vapour followed by carbon dioxide (CO_2). There are smaller contributions from many other gases including ozone, methane, nitrous oxide and human-made gases such as CFCs (chlorofluorocarbons).

Climate change

Climate change on a global scale, whether natural or due to human activity, can be initiated by processes that modify either the amount of energy absorbed from the Sun, or the amount of infrared energy emitted to space.

Climate change can therefore be initiated by changes in the energy received from the Sun, changes in the amounts or characteristics of greenhouse gases, particles and clouds, or changes in the reflectivity of the Earth's surface. The imbalance between the absorbed and emitted radiation that results from these changes will be referred to here as 'climate forcing' (sometimes known as 'radiative forcing') and given in units of W/m^2. A positive climate forcing will tend to cause a warming, and a negative forcing a cooling. Climate changes act to restore the balance between the energy absorbed from the Sun and the infrared energy emitted into space.

> **Human activity is a relatively recent addition to the list of potential causes of climate change**

In principle, changes in climate on a wide range of time-scales can also arise from variations within the climate system due to, for example, interactions between the oceans and the atmosphere; in this article, this is referred to as 'internal climate variability'. Such internal variability can occur because the climate is an example of a chaotic system: one that can exhibit complex unpredictable internal variations even in the absence of the climate forcings discussed in the previous paragraph.

There is very strong evidence to indicate that climate change has occurred on a wide range of different timescales, from decades to many millions of years; human activity is a relatively recent addition to the list of potential causes of climate change.

The shifts between glacial and interglacial periods over the past few million years are thought to have been a response to changes in the characteristics of the Earth's orbit around the Sun. While these led to only small changes in the total energy received from the Sun, they led to significant changes in its geographical and seasonal distribution. The large changes in climate, in moving in and out of glacial periods, provide evidence of the sensitivity of climate to changes in the Earth's energy balance, whether attributable to natural causes or to human activity.

Mechanisms of global climate change

Once a climate forcing mechanism has initiated a climate response, this climate change can lead to further changes: for example, in response to a warming, the amount of water vapour is expected to increase, the extent of snow and ice is expected to decrease, and the amount and properties of clouds could also change. Such changes can further modify the amount of energy absorbed from the Sun, or the amount of energy emitted by the Earth and its atmosphere, and lead to either a reduction or amplification of climate change.

The overall effect of the changes resulting from climate forcing determine a key characteristic of the climate system, known as the 'climate sensitivity' – this is the amount of climate change (as measured by the equilibrium change in globally-averaged surface temperature) caused by a given amount of climate forcing. It is often quoted (as will be the case here) as the temperature change that eventually results from a doubling in CO_2 concentrations since pre-industrial times, and is calculated to cause a climate forcing of about 3.6 W/m^2.

The nature of the climate system is determined by interactions between the moving atmosphere and oceans, the land surface, the living world and the frozen world. The rate at which heat is moved from the surface to the ocean depths is an important factor in determining the speed at which climate can change in response to climate forcing.

Since variations in climate can result from both climate forcing and internal climate variability, the detection of forced climate change in observations is not always straightforward. Furthermore, the detection of climate change in observations, beyond the expected internal climate variability, is not the same as the attribution of that change to a particular cause or causes. Attribution requires additional evidence to provide a quantitative link between the proposed cause and the observed climate change.

Modelling the climate system

Current understanding of the physics (and increasingly the chemistry and biology) of the climate system is represented in a mathematical form in climate models, which are used to simulate past climate and provide projections of possible future climate change. Climate models are also used to provide quantitative estimates to assist the attribution of observed climate change to a particular cause or causes.

Climate models vary considerably in complexity. The simplest can be described by a few equations, and may represent the climate by global-average surface temperature alone. The most complicated and computer-intensive models represent many details of the interactions between components of the climate system. These more complex models represent variations in parameters such as temperature, wind and humidity with latitude, longitude and altitude in the atmosphere, and also represent similar variations in the ocean. In complex climate models climate sensitivity emerges as an output; in the simpler calculations it is specified either as an input or it emerges as a consequence of simplified (but plausible) assumptions.

By applying established laws of fluid dynamics and thermodynamics, the more complex climate models simulate many important weather phenomena that determine the climate. However, limitations of computer power mean that these models cannot directly represent phenomena occurring at small scales. For example, individual clouds are represented by more approximate methods. Since there are various ways to make these approximations, the representation can vary in climate models developed at different climate institutes. The use of these different approximations leads to a range of estimates of climate sensitivity, especially because of differences between models in the response of clouds to climate change. There are intensive efforts to compare the models with observations and with each other. The spread of results from these models gives useful information on the degree of confidence in the reliability of projections of climate change.

Unlike weather-forecast models, climate models do not seek to predict the actual weather on a particular day at a particular location. The more complex models do however simulate individual weather phenomena, such as mid-latitude depressions and anticyclones, and aim to give simulations of possible weather sequences much farther into the future than weather-forecast models. From such simulations, one can derive the characteristics of climate likely to occur in future decades, including mean temperature and temperature extremes.

September 2010

⇨ This is an extract from The Royal Society's *Climate change: a summary of the science*, which summarises the evidence and clarifies the levels of confidence associated with the current scientific understanding of climate change. It can be downloaded and read in full at http://royalsociety.org/Climate-Change

© The Royal Society

Myths about climate change

Confusion and myths about climate change are widespread. Explore some of the most common misconceptions and the facts behind them.

The climate is always changing anyway

The Earth's climate has always changed naturally in the past. But what is happening now is potentially a big change in the Earth's climate, this time caused mainly by human activity.

Carbon dioxide is a major heat-trapping greenhouse gas. Its concentration in the atmosphere is now higher than at any time in at least the last 800,000 years. Although this is not new in the history of the planet, it is entirely new in human history. It is expected to have a negative impact on many ecosystems and humans across the world.

There's no scientific evidence for climate change

Scientists have been commenting on the relationship between emissions of gases and the climate since the 1800s. They have worked with governments to do something about climate change for a long time.

In 1988, the UN set up the Intergovernmental Panel on Climate Change (IPCC). The IPCC is a body of scientists from all parts of the world who assess the best available scientific and technical information on climate change.

The IPCC's 2007 Fourth Assessment Report warned of a rise in average global temperatures. This rise could be from 1.1 to 6.4 degrees Celsius above 1980-1999 levels by the end of this century, depending on future levels of emissions. Based on current science, the report said that recent temperature increases were very likely (over 90 per cent probable) the result of human activity.

Climate change isn't caused by human activity

Based on a vast amount of evidence, nearly all climate science experts are convinced that humans are affecting the climate by the way they live. The Met Office Hadley Centre is one of the world's leading centres for climate change research. It found that recent temperature rises and key changes in the Earth's environment could not be explained by natural climate change alone. Human activity is mainly responsible.

Scientific research and careful observation has shown that the concentration of greenhouse gases, which keep the Earth warm, is increasing. People are responsible for these increases by, for example, burning fossil fuels and cutting down forests.

Over 40 per cent of CO_2 emissions in the UK come directly from what people do – for example, using electricity in the home and driving cars. If every home installed the recommended amount of loft insulation, it would save 3.8 tonnes of CO_2. This is the same as the emissions of about 650,000 homes in one year.

It's too late to make a difference

The last report from the IPCC in 2007 said that, if the world is to avoid dangerous climate change, global greenhouse gas emissions must:

⇨ peak within the next decade or two;

⇨ decline rapidly to well below current levels by the middle of the century.

This is still possible, and may be achieved with technologies that are available now. Putting off action to cut greenhouse gases will make it increasingly difficult and expensive to reduce emissions in the future. It will also create higher risks of severe climate change impacts.

There's no point in me taking action

Every reduction in emissions makes a difference by not adding to the risk. Countries like the UK are in a position to give a positive example to the rest of the world. If the UK can rise to the challenge successfully, others will follow.

Cutting my carbon footprint will affect my lifestyle

There are many small and simple things you can do that will contribute to big reductions in carbon emissions. Many actions will have little to no effect on your lifestyle – for example:

⇨ turning off the lights when you leave a room;

⇨ switching appliances off at the mains;

⇨ turning your thermostat down one degree.

Climate change will make life more comfortable in the UK

Climate change will lead to warmer winters, but temperatures will become uncomfortably hot in summer,

and the climate may also be unpredictable and extreme. There's also the risk of rising sea levels and extreme weather like storms and floods. Tackling climate change and securing a more stable climate will make life a lot more comfortable.

It would cost too much to tackle climate change

Tackling climate change needn't damage the economy as a whole. Industry will have to adapt and jobs may change – but more may be created overall. Using less energy can also save companies and households money.

Climate change will lead to warmer winters, but temperatures will become uncomfortably hot in summer

Not tackling climate change has a price too. The recent Stern report examines the economic impact of climate change. It estimates that not taking action could cost from five to 20 per cent of global GDP (gross domestic product) every year, now and forever. In comparison, reducing emissions to avoid the worst impacts of climate change could cost around one per cent of global GDP each year.

⇨ The above information is reprinted with kind permission from DirectGov. Please visit www.direct.gov.uk for more information.

© Crown copyright

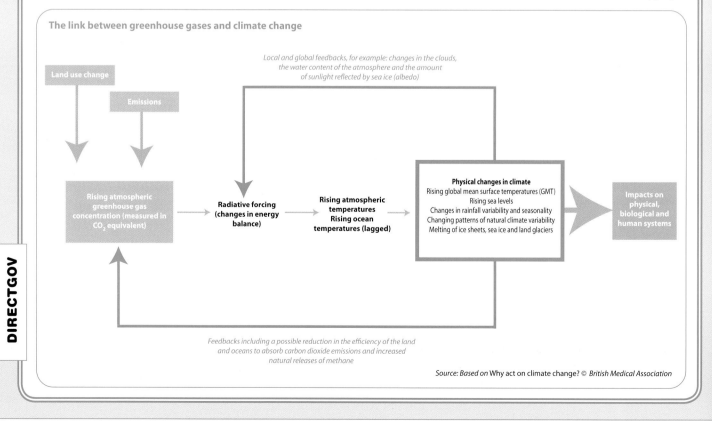

The link between greenhouse gases and climate change

Local and global feedbacks, for example: changes in the clouds, the water content of the atmosphere and the amount of sunlight reflected by sea ice (albedo)

Land use change

Emissions

Rising atmospheric greenhouse gas concentration (measured in CO_2 equivalent)

Radiative forcing (changes in energy balance)

Rising atmospheric temperatures
Rising ocean temperatures (lagged)

Physical changes in climate
Rising global mean surface temperatures (GMT)
Rising sea levels
Changes in rainfall variability and seasonality
Changing patterns of natural climate variability
Melting of ice sheets, sea ice and land glaciers

Impacts on physical, biological and human systems

Feedbacks including a possible reduction in the efficiency of the land and oceans to absorb carbon dioxide emissions and increased natural releases of methane

Source: Based on Why act on climate change? *© British Medical Association*

DIRECTGOV

The status of climate change science today

A factsheet from the United Nations Framework Convention on Climate Change.

The World Meteorological Organization (WMO) describes the build-up of greenhouse gases in the atmosphere during the 20th century as resulting from the growing use of energy and expansion of the global economy. According to the WMO, the build-up of greenhouse gases in the atmosphere alters the radiative balance of the atmosphere. The net effect is to warm the Earth's surface and the lower atmosphere because greenhouse gases absorb some of the Earth's outgoing heat radiation and reradiate it back towards the surface.

> *The build-up of greenhouse gases in the atmosphere alters the radiative balance of the atmosphere*

The most recent comprehensive assessment of the science was undertaken in 2007 by the Intergovernmental Panel on Climate Change (IPCC) on the causes, impacts and possible response strategies to climate change. The conclusions are supported by a wide range of the world's leading scientific institutions, including the US's National Oceanic and Atmospheric Administration (NOAA). During 2010, there has been widespread debate about climate science, particularly as a result of errors which emerged in the last (2007 AR4) IPCC report. None of the errors alter the fundamental conclusions of the IPCC's AR4, namely that climate change is the result of human activity, that the phenomenon will have devastating effects if left unchecked and that costs of action on climate change are significantly lower than the costs of inaction. Following a review by the InterAcademy Council (IAC), the IPCC has announced that it will strengthen a number of its processes and procedures. The IPCC Fifth Assessment Report (AR5) will be published between 2013 and 2014.

During 2010, several regions of the world experienced what the WMO terms severe weather-related events. These included flash floods and widespread flooding in large parts of Asia and parts of Central Europe. Other regions were also affected: by heatwave and drought in the Russian Federation, by mudslides in China and severe droughts in sub-Saharan Africa. The WMO stated that while a longer time range is required to establish whether an individual event is attributable to climate change, the sequence of current events matches IPCC projections of more frequent and more intense extreme weather events due to global warming.

The IPCC Fourth Assessment Report

The IPCC Fourth Assessment Report, *Climate change 2007: Synthesis Report*, comprises contributions from the three working groups on (1) the physical science, (2) climate change impacts, adaptation and vulnerability and (3) mitigation of climate.

Definitions of climate change

Climate change in IPCC usage refers to a change in the state of the climate that can be identified (e.g. using statistical tests) by changes in the mean and/or the variability of its properties, and that persists for an extended period, typically decades or longer. It refers to any change in climate over time, whether due to natural variability or as a result of human activity.

This usage differs from that in the United Nations Framework Convention on Climate Change (UNFCCC), where climate change refers to a change of climate that is attributed directly or indirectly to human activity, that alters the composition of the global atmosphere, and that is in addition to natural climate variability observed over comparable time periods.

The physical science

⇨ Warming of the climate system is unequivocal and can now be firmly attributed to human activity.

⇨ Numerous long-term changes in climate have been observed at continental, regional and ocean basin scales, including changes in arctic temperatures and ice, widespread changes in precipitation amounts, ocean salinity, wind patterns and aspects of extreme weather including droughts, heavy precipitation, heat waves and the intensity of tropical cyclones.

⇨ The 100-year linear warming trend (1906-2005) was 0.74°C, with most of the warming occurring in the past 50 years. The warming for the next 20 years is projected to be about 0.2°C per decade.

⇨ Continued greenhouse gas emissions at or above current rates would cause further warming and

UNFCC

induce many changes in the global climate system during the 21st century that would very likely be larger than those observed during the 20th century.

Hot extremes, heat waves and heavy precipitation events are very likely to continue and become more frequent

⇨ Projections of future changes in climate indicate, for example, the following:

↳ Increasing atmospheric carbon dioxide concentrations lead to increasing acidification of the oceans;

↳ Snow cover projected to contract, widespread increases in thaw depth over most permafrost regions;

↳ Sea ice projected to shrink in both Arctic and Antarctic, and Arctic late-summer sea ice may disappear almost entirely by the latter part of the 21st century;

↳ Hot extremes, heat waves and heavy precipitation events are very likely to continue and become more frequent;

↳ Changes in precipitation patterns, with increases being very likely in high latitudes, while decreases are likely in most subtropical land regions;

↳ Anthropogenic warming and sea-level rise would continue for centuries due to the time-scales associated with climate processes and feedbacks, even if greenhouse gas concentrations were to be stabilised.

Climate change impacts, adaptation and vulnerability

Observed impacts

Many natural systems, on all continents and most oceans, are being affected by regional climate changes, particularly temperature increases. Observed impacts include:

⇨ Changes in snow, ice and frozen ground (including permafrost).

⇨ Effects on hydrological systems.

⇨ Changes on terrestrial biological systems.

⇨ Trend towards earlier 'greening' of vegetation and longer thermal growing season.

⇨ Changes in marine and freshwater biological systems associated with rising water temperatures, as well as related changes in ice cover, salinity, oxygen levels and circulation.

⇨ Ocean acidification with an average decrease in pH of 0.1 units. The associated effects on the marine biosphere were not documented at the time of the assessment.

Projected future impacts

As regards the projected impacts, more specific information is now available on the nature of these impacts, across a wide range of systems and sectors. Examples of projected impacts include:

Freshwater resources and their management

⇨ Runoff and water availability are projected to increase at high latitudes and in some wet tropics, and decrease over much of the mid-latitudes and dry tropics, some of which are presently water-stressed areas.

⇨ Drought-affected areas will probably increase, and extreme precipitation events, which are likely to increase in frequency and intensity, will augment flood risk.

⇨ Hundreds of millions of people are projected to be exposed to increased water stress.

Ecosystems

⇨ The following ecosystems are identified to be most vulnerable, and are virtually certain to experience the most severe ecological impacts, including species extinctions and major biome changes:

↳ On continents: tundra, boreal forest, mountain and Mediterranean-type ecosystems.

↳ Along coasts: mangroves and salt marshes, due to multiple stresses.

↳ In oceans: coral reefs and the sea-ice biomes.

⇨ The progressive acidification of the oceans is expected to have negative impacts on marine shell-forming organisms such as corals and their dependent species.

⇨ An intensification and expansion of wildfires is likely globally, as temperatures increase and dry spells become more frequent and more persistent.

⇨ Over the course of this century, net carbon uptake by terrestrial ecosystems is likely to peak before mid-century and then weaken or even reverse, thus amplifying climate change.

Food, fibre and forest products

⇨ Moderate warming benefits cereal crops and pasture yields in mid- to high-latitude regions, but even slight warming decreases yields in seasonally dry and tropical regions. Further warming has increasingly negative impacts in all regions.

- Increases in the frequency of droughts and floods are projected to affect local crop production negatively, especially in subsistence sectors at low latitudes.

- Regional changes in the distribution and production of particular fish species are expected due to continued warming, with adverse effects projected for aquaculture and fisheries.

Coastal areas and low-lying areas

- Coasts are very likely to be exposed to increasing risks, including coastal erosion, due to climate change and sea-level rise. The effect will be exacerbated by increasing human-induced pressures on coastal areas.

- In addition to sea-level rise, low-lying coastal systems are likely to be affected due to increased risk from extreme weather events.

- Many millions more people are projected to experience severe flooding every year due to sea-level rise by the 2080s. Those densely-populated and low-lying areas where adaptive capacity is relatively low, and which already face other challenges such as tropical storms or local coastal subsidence, are especially at risk. The numbers affected will be largest in the mega-deltas of Asia and Africa, while small islands are especially vulnerable.

- Ocean acidification is an emerging issue with potential for major impacts in coastal areas, but there is little understanding of the details. It is an urgent topic for further research.

Coasts are very likely to be exposed to increasing risks, including coastal erosion

Health

- Projected climate change-related exposures are likely to affect the health status of millions of people worldwide, particularly those least able to adapt, such as the poor, the very young and the elderly.

Industry, settlement and society

- Areas most likely to be affected are the poorer, often rapidly expanding communities near rivers and coasts, which use climate-sensitive resources and are prone to extreme weather.

- Where extreme weather events become more intense and/or more frequent, their economic and social costs are predicted to increase.

Regions that will be especially affected

- The Arctic, due to impacts of high rates of projected warming on natural systems and human communities.

- Africa, because of low adaptive capacity and projected climate change impacts.

- Small islands, where there is high exposure of population and infrastructure to projected climate change impacts.

- Asian and African megadeltas, due to large populations and high exposure to sea level rise, storm surges and river flooding.

Mitigation of climate change

GHG emission trends

- Global greenhouse gas emissions have grown since pre-industrial times, with an increase of 70 per cent between 1970 and 2004 (24 per cent between 1990 and 2004).

- With current climate change mitigation policies and related sustainable development practices, global GHG emissions will continue to grow over the next few decades.

Mitigation in the short and medium term up to 2030

There is a substantial economic potential for the mitigation of global greenhouse gas emissions over the coming decades, sufficient to offset the projected growth of global emissions or reduce emissions below current levels.

Mitigation in the long term (after 2030)

- Global emissions must peak and decline thereafter to meet any long-term GHG concentration stabilisation level.

- The lower the stabilisation level, the more quickly this peak and decline must occur.

- The most stringent scenarios could limit global mean temperature increases to 2-2.4°C above pre-industrial level. This would require emissions to peak by 2015 at the latest and decline by 50-85 per cent compared to year 2000 emissions by 2050.

- Mitigation efforts over the next two to three decades will determine to a large extent the long-term global mean temperature increase and the corresponding climate change impacts that can be avoided.

February 2011

- The above information is reprinted with kind permission from the United Nations Framework Convention on Climate Change. Visit www.unfccc.int for more information.

UNFCC

Abrupt climate change

Information from the David Suzuki Foundation.

Recent climate change research has uncovered a disturbing feature of the Earth's climate system: it is capable of sudden, violent shifts. This is a critically important realisation. Climate change will not necessarily be gradual, as assumed in most climate change projections, but may instead involve sudden jumps between very different states. This would present an enormous challenge to human societies and the global environment as a whole.

While no such violent shift has occurred over the duration of human civilisation, records of prehistoric shifts are clear. By studying a range of long-term natural records – including tree rings, cave deposits, ice cores and deep-sea sediments – scientists have begun to assemble an almanac of global change spanning hundreds of thousands of years. One of the most important findings to come from these records is that the climate system has undergone many sudden shifts.

The Great Ocean Conveyor

The Great Ocean Conveyor (also called the thermohaline circulation) is one of the great unknowns in the climate system. Surface water, warmed at the equator, circulates via the Gulf Stream to high latitudes where it releases heat to the atmosphere. As a result, the water cools, becoming denser, and sinks to the deep ocean.

Computer models and paleoclimate evidence both suggest that increases in freshwater supply to the North Atlantic ocean – for example, from increased rainfall or melting glaciers – can lower the salinity of the surface waters and make them too buoyant to sink, disrupting the ocean's circulation. As a result, the northward transport of heat stops abruptly, and temperatures around the North Atlantic drop.

A distressing confirmation of the models has recently emerged. Measurements of the saltiness of the North Atlantic show that the region's waters have been growing gradually fresher over the past 40 years. The rapidity and extent of freshening came as a surprise to oceanographers. The change is equivalent to a three- to four-metre cap of fresh water appearing over a broad area of the northern North Atlantic. If this continues, the resulting loss of density could prevent sinking, and short-circuit the Conveyor within decades.

Global warming may cause severe local cooling

It is well accepted that the Earth as a whole is warming as a result of human activity. Ironically, the lessons of distant Earth history show that warming may cause a sudden drop in temperature in some regions. The heat released to the atmosphere by the North Atlantic loop of the Conveyor is largely responsible for the relatively warm temperatures enjoyed by Western Europe. If not for this, European winters would be much colder. Berlin might have the climate of Edmonton, Canada, which lies at the same latitude, while Stockholm might be more like Iqaluit.

Judging from records of the distant past, this could be disastrous for eastern North America and Western Europe. Average temperatures could plunge by five degrees Celsius. This is about the same difference as between the global average temperature today and during the last Ice Age, when Canada lay beneath 3,000-metre-thick glaciers. The resulting colder, snowier winters would require new infrastructure, damage crops and shorten the growing season. The costs associated with such a change would be enormous.

⇨ The above information is reprinted with kind permission from the David Suzuki Foundation. Visit www.davidsuzuki.org for more information.

DAVID SUZUKI FOUNDATION

What about climate change in the future?

We simply don't know exactly what the UK's climate will do very long-term.

How our climate will change depends on the future level of carbon dioxide and other gas emissions in the atmosphere. Some impacts are also highly unpredictable in a complex climatic system.

The experts believe there is no likelihood of the Gulf Stream closing down within the next two decades, but it is a possibility longer term.

So while we look set for a warmer – and stormier – climate for at least the next 20 to 30 years, very long-term who knows?

How our climate will change depends on the future level of carbon dioxide and other gas emissions in the atmosphere

On the best projections now available from the International Climate Change Panel and UK Climate Impacts Programme, these are some of the anticipated changes over 75 years:

Temperature

⇨ Globally temperatures could rise anywhere between 1.5 and 5.8°C by 2080 – between twice and eight times the rise we have already seen since 1900.

⇨ Each degree of warming causes a lengthening of the growing season in the UK by between 1.5 weeks in the north and three weeks in the south.

⇨ In the UK, an average temperature rise of 2-3.5°C is anticipated by 2080; though some areas could warm by nearly 6°C.

⇨ More heat waves in summer are predicted – perhaps what we would now call an exceptional summer, like 1995, occurring two years out of three by 2080.

Rain and snow

⇨ Winters will become wetter (20-35 per cent wetter by 2080) and summers may become drier (35-50 per cent drier by 2080).

⇨ Snowfalls will become increasingly rare – maybe up to 90 per cent less snow by 2080.

⇨ Heavier rainfall events will become more frequent; though this cannot be quantified.

⇨ Up to 50 per cent reduction in soil moisture content in south and east by 2080.

The sea

(This is heavily dependent on the speed of melting of Polar ice caps.)

⇨ Extreme sea levels could occur between ten and 20 times more frequently by the 2080s.

⇨ Relative sea level will continue to rise around most of the UK, perhaps by as much as 86 cm in south-east England.

⇨ Sea temperatures are expected to continue to rise over the next couple of decades, though more slowly than the land temperature rises.

⇨ The ranges shown here reflect the fact that we don't know future levels of carbon emissions which then determine climate.

⇨ The above information is reprinted with kind permission from the National Trust. Visit www.nationaltrust.org.uk for more information.

© National Trust

The social and economic impacts of climate change

south essex college
FURTHER & HIGHER EDUCATION
SOUTHEND CAMPUS

Information from Earthwatch.

Social impacts

Climate change is likely to have far-reaching and catastrophic social impacts and will affect communities in different ways. Vulnerability to climate change impacts depends on differences in geography, technological resources, governance and wealth. It is often the world's most impoverished nations that are most vulnerable to the effects of climate change – nations that have the fewest resources to adapt and cope with these effects. Communities in developing countries which are making little or no direct contribution to climate change are likely to be among the most affected.

Water shortages

Declining rainfall and accelerated evaporation may reduce runoff, threatening the availability of fresh water for human and industrial consumption. Furthermore, loss of glaciers and ice fields may jeopardise drinking water supplies. Water supply is likely to be affected both in terms of quantity and in quality as demand for depleting supplies increases. Loss of water (and food) security may lead to increased conflict. In Kenya, for example, there have been territorial disputes over receding water bodies, and increases in cattle raiding and violence, as people who have historically managed through periods of drought and food shortages find themselves dealing with unprecedented famine.

Food shortages

Climate change will affect food production, as well as how much food is available to people (food security). Increases in temperature in high latitudes will extend the growing range of some agricultural crops, changing seasons will affect the growing seasons, and increased atmospheric CO_2 will boost agricultural productivity.

These potential benefits, however, will be matched by altered weather patterns which will increase crop vulnerabilities to infection, pest infestations and choking weeds. This will not only decrease yields of crops, but also force farmers to increase application of harmful and expensive pesticides and herbicides. The increase in extreme weather events will affect both developed and developing countries, although developed countries have more resources to deal with vulnerabilities. Impoverished, marginalised people are often directly dependent on the diversity in local ecosystems to support their livelihoods. If climate change is not averted,

an additional 80 to 120 million people will be at risk of hunger. 70 to 80% of these people will be in Africa, and the majority are likely to be women, who have a greater reliance on subsistence farming.

Health

The health implications of climate change are profound. Climate change will increase experience of heat stress, injury and death from natural disasters (such as floods and windstorms), vector-borne diseases (such as malaria, dengue, schistosomiasis and tick-borne diseases), water- and food-borne diseases. The elderly and women are likely to be disproportionately affected by the increased disease burden. In developing countries like Africa, where severe health problems such as malaria, HIV/AIDS and hunger-related diseases are already widespread, the added health implications of climate change are likely to result in an increase in human mortality. Rising temperatures may increase risks associated with aquatic pathogens in important fisheries, and accelerate the spoiling of food and meat.

Climate change refugees

It is estimated that by 2050 there will be 250 million people who will be forced to flee their homes due to drought, desertification, sea-level rise and extreme weather events. Many human populations on islands in the Pacific have already become victims of climate change. For example, many people living on the Carteret islands of Papua New Guinea have been evacuated due to rising sea levels making their islands uninhabitable.

Given the rapid urbanisation of the world's developing countries, the impact of climate change on the housing and wellbeing of urban populations is especially important. Like their rural counterparts, the most vulnerable urban populations will bear the greatest impacts. The major long-term impacts are likely to be severe housing shortages and overcrowding as rural populations are displaced by drought and flooding. The lack of safe water and sanitary infrastructure in emergency camps or slum areas could seriously increase the incidence of mortality and morbidity from transmissible diseases. The Intergovernmental Panel on Climate Change reported in 2001 that these internal environmental refugees may 'present the most serious health consequences of climate change'. Violence and social tension, already severe in the poorest slums and shanty towns of places such as Latin American and Caribbean cities, are likely to intensify.

Those displaced by natural disasters in rural areas may remain at risk in urban areas where shanty towns

and slums are often situated on land prone to flooding or landslide. Increased intensity and frequency of extreme weather events will threaten these precarious settlements and their marginalised populations.

Implications on quality of life

Biodiversity has long been recognised as a contributing factor to quality of life. In 2011, the UK Government's Department of Environment, Farming and Rural Affairs (DEFRA) carried out a survey on attitudes and knowledge relating to biodiversity and the natural environment. It showed that:

⇨ 63% of respondents gave at least a little thought to the loss of biodiversity in the UK, compared to 46% in 2009;

⇨ 78% of respondents agreed that they 'worry about the changes to the countryside in the UK and loss of native animals and plants';

⇨ 85% of respondents agreed that they were 'proud of their local environment'.

Economic impacts

Agricultural production

Changes in climate will impact on agriculture and food production in many ways. Production may increase with higher temperatures in middle and higher latitudes, since the length of the potential growing season may be increased. Crop-producing areas may expand poleward in countries such as Canada and Russia, and in some regions new opportunities will occur for growing new crops such as grapes for wine.

Increasing levels of CO_2 will impact on different crops in different ways, by affecting the rate of photosynthesis. Experiments have demonstrated that crops such as wheat, rice and soybeans respond positively to increased CO_2 by increased growth, whereas other crops such as sugarcane, millet and corn are less responsive.

Agriculture of any kind is strongly influenced by the availability of water. The demand for water for irrigation is projected to rise in a warmer climate. Falling water tables and the resulting increase in the energy needed to pump water will make the practice of irrigation more expensive, particularly when, with drier conditions, more water will be required.

Agricultural pests may increase in warmer climates. Longer growing seasons will enable insects such as grasshoppers to complete a greater number of reproductive cycles during the spring, summer and autumn. Warmer winter temperatures may also allow larvae to over-winter in areas where they are now limited by cold, thus causing greater infestation during the following crop season.

Developing countries' economies and climate change

The economies of developing countries are highly dependent on agriculture and ecosystems, and changes in agricultural output will undoubtedly affect these nations. 61% of people in South Asia and 64% in sub-Saharan Africa are employed in the rural agricultural sector, and the agricultural sector is the most at risk from climate change (*Stern Review: The Economics of Climate Change*).

The economic strain of coping with the many social impacts of climate change on countries (and especially developing countries) such as reduced food security, increase of extreme weather events such as flooding and increased diseases such as malaria, will be great. For example, Zimbabwe suffered from an extreme drought from 1991 to 1992 which caused GDP (Gross Domestic Product) to fall by 9% and inflation to increase by 46% (*Stern Review: The Economics of Climate Change*). Government spending on education and health was sacrificed for drought-related emergency outlays. An adverse climate may mean that people or governments coping with its effects will be locked into a poverty trap as the majority of their income and assets are spent on coping strategies rather than profit-making strategies.

Forestry

A report compiled by the Forestry Commission predicted that under various climate change predictions, the growing distribution and productivity of different commercially-grown tree species will be altered. For example, Sitka Spruce will be unable to be grown at low altitudes, but will show increased productivity at higher altitudes in the UK, whereas future climates will decrease the range and productivity of Scot's Pine across the whole of the UK. In addition, an increase in the frequency of forest fires due to more droughts and less precipitation in some areas will affect the forestry industry.

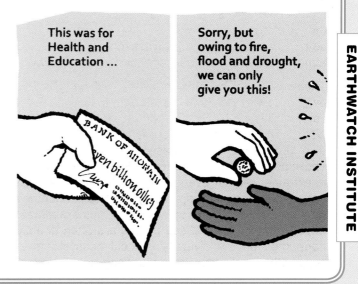

Tourism

Tourism is one of the fastest growing industries in the European Union and the most important industry in many developing countries. Changing climates and weather patterns will alter the amount of tourism to certain countries, since the appeal of tourism relies heavily on the natural environment.

While tourism may become a victim of climate change, it also contributes to climate change through air travel. Air travel accounts for around 3% of CO_2 emissions and the International Panel on Climate Change expects this to increase to 7% by 2050.

Some destinations such as the Maldives, which are vulnerable to sea-level rise, could see reduced numbers of tourists, while the Mediterranean will experience an increased number of extreme heat days (above 40°C) which could deter tourists. Winter tourism and the skiing industry are under threat due to shorter ski seasons and unpredictable snow conditions. In the European Alps, skiing represents about 5% of Alpine countries' GDP; therefore, the threat of loss of skiing and winter sports tourism could impact heavily on these countries' economies.

⇨ The above information is reprinted with kind permission from the Earthwatch Institute. Visit www. earthwatch.org for more information.

© Earthwatch Institute

West Antarctic Ice Sheet 'could be more stable than thought'

Whether global warming may cause the giant West Antarctic Ice Sheet to melt and raise sea-levels by several metres is one of the most contested debates in climate science.

A new study involving a scientist at the University of Exeter has reported valuable new evidence on the issue which suggests the ice sheet may be more stable than thought.

Dr Chris Fogwill, from the University's Geography department in the College of Life and Environmental Sciences, went to Antarctica to carry out first-hand research for the Natural Environment Research Council (NERC)-funded study.

He said: 'The debate on the ice sheet focuses around the Earth's past climate. Evidence suggests our climate has warmed before and about 125,000 years ago there was an "interglacial period" when global temperatures were about 2°C warmer than they are today.

'Some models of that past climate suggest sea-levels were much higher during that time than they are now, and some of that water would have to have come from this giant freshwater body of ice – suggesting the ice sheet is vulnerable to melting at warmer global temperatures.

'However, we found evidence which suggests the ice sheet has been around for at least 200,000 years, meaning that it has survived at least one warm period and is more resilient than thought.'

Dr Fogwill, working with a team from the University of Edinburgh, used a state of the art method called cosmogenic dating which relies on radiation left over from the Big Bang – the cataclysmic event which scientists believe created the Universe.

Cosmic rays build up in rock when it is exposed. Dr Fogwill looked at the cosmic radiation levels in exposed rocks in moraines, where weather and landscape erode ice to reveal bare rock, along the Heritage range of mountains near the central dome of the West Antarctic Ice Sheet.

They found that the moraines had been developing for at least 200,000 years, suggesting ice has covered the area for at least that long – meaning the ice sheet would have survived the last warm period in the Earth's climate.

Dr Fogwill added: 'This research doesn't provide conclusive evidence, but it definitely provides us with a solid theory. There is evidence from other studies which suggests the ice sheet isn't as resilient, so this will remain an area of uncertainty for now.'

However, Dr Fogwill has just returned from a recent trip to the Antarctic to gather more evidence to study to test this hypothesis. He has collected more samples for cosmogenic dating and will be collating the information to test the theory from the newly published study.

The full paper, published in *Palaeogeography, Palaeoclimatology, Palaeoecology,* can be viewed online.

11 March 2011

⇨ Information from the University of Exeter. Visit www.exeter.ac.uk for more.

© University of Exeter

Top ten global weather/climate events of 2010

Top ten list from the National Oceanic and Atmospheric Administration, National Climatic Data Center.

The following article lists the top ten global weather/climate events of 2010. These events are listed according to their overall rank, as voted on by a panel of weather/climate experts. The voters considered factors such as the scope and unusualness of the event, its immediate human and economic impact, and whether it is emblematic of climate trends or variability.

Note: These lists were compiled and voted on during the first week of December. Significant events, such as the extreme winter weather in Europe and the flooding in Australia, occurred after this date. These events may have warranted top ten placement.

1 *Russian – European – Asian Heat Waves – Summer*

A severe summer spawned drought, wildfires and crop failures across western Russia, where more than 15,000 people died. All-time high temperatures occurred in many cities and nations in the region. China faced locust swarms during July.

2 *2010 as [near] warmest year on record – calendar year*

According to NOAA, the globally-averaged temperature for 2010 will finish among the two warmest, and likely the warmest, on record. Three months in 2010 were the warmest on record for that month.

3 *Pakistani Flooding – Late July-August*

Rainfall related to the Asian Monsoon was displaced unusually westward, and more than a foot of rain fell across a large area of the Upper Indus Valley. Subsequent flooding down the Indus River killed 1,600 people and displaced millions.

4 *El Niño to La Niña Transition – Mid-to-Late Boreal Spring*

El Niño Southern Oscillation, the most prominent and far-reaching patterns of climate variability, saw a huge swing in mid-2010. Only 1973, 1983 and 1998 have seen larger within-year swings.

5 *Negative Arctic Oscillation – Early 2010*

The AO Index, which is strongly correlated with wintertime cold air outbreaks, reached -4.27 for February, the largest negative anomaly since records began in 1950. Major cold air outbreaks occurred throughout the Northern Hemisphere.

6 *Brazilian Drought – Ongoing*

A severe drought parching northern Brazil shrunk the Rio Negro, one of the Amazon River's most important tributaries, to its lowest level since records began in 1902, at its confluence with the Amazon. The Amazon's depth there fell more than 12 feet below its average.

7-tie *Historically Inactive NE Pacific Hurricane Season – 15 May-30 Nov*

The Northeast Pacific Hurricane Season was one of the least active on record, produced the fewest named storms and hurricanes of the modern era, and had the earliest cessation of tropical activity (23 September) on record.

7-tie *Historic N. Hemispheric Snow Retreat – January-June*

Despite December 2009 having the second-largest snow cover extent of the satellite record (mid-1960s), the melt season was ferocious, contributing to spring floods in the northern US and Canada. Following the early and pronounced snow melt, the North American, Eurasian and Hemispheric snow cover was the smallest on record for May and June 2010.

9 *Minimum Sea Ice Extent – mid-September*

The 2010 sea ice minimum of 4.9 million sq km was the third smallest on record. The last four years (2007-2010) are the four smallest on record. The Northwest Passage and the Northern Sea Route were simultaneously ice-free in September, a first in modern history.

10 *China Drought – First Half of 2010*

A persistent drought centred in the Yunan Province was touted as perhaps the worst in this region in more than 100 years. Major crop losses and lack of drinking water created severe problems for local residents.

⇨ The above information is reprinted with kind permission from National Climatic Data Center (NOAA). Visit www.ncdc.noaa.gov for more information.

© National Climatic Data Center (NOAA)

NATIONAL CLIMATIC DATA CENTER (NOAA)

Climate change: the forest connection

Most people are now aware that the world's hunger for energy from fossil fuel is leading to catastrophic climate change. What is also becoming increasingly clear is the effect that forests have on the climate and the climate has on forests – and how changes in one system will affect the other.

Forests' effect on the climate

Forests play an important role in regulating the Earth's temperature and weather patterns by storing large quantities of carbon and water. This regulatory function has a profound effect on both the local and the global climate.

Locally, trees provide shade, which in turn lowers summer temperatures and prevents the soil from drying out. They reduce heat loss from the ground in winter and reduce storm damage by providing shelter from wind.

Globally, forests regulate the global carbon cycle, having a profound effect on the climate. As well as this, deforestation is also contributing to climate change. Indeed, the CO_2 released each year from forest loss is higher than that released by our yearly transport emissions. The continued existence of forests is particularly necessary if we are to halt what is known as runaway climate change. Runaway climate change is the point whereby increases in temperature lead to more GHG emissions which in turn leads to increased temperatures. Examples of this include: increased temperatures will melt ice caps which will lead to huge increases in the release of GHGs; and increases in temperature are projected to negatively affect up to two-thirds of existing forests, thereby exacerbating deforestation and increasing the release of carbon.

The climate's effect on forests

Global warming, which on a geological timescale is occurring in the equivalent of a split second, is significantly disrupting the intricate and poorly understood web of interactions that governs the very structure and composition of forest ecosystems. This means that around a third of today's forests are likely to change their species composition. A temperature increase of 3°C by 2100 would result in forest ecosystems having to move 500 km towards the poles or 500 m in elevation in order to find the same climatic conditions. Such distances are far beyond the average rate of dispersal for individual tree species, let alone entire forest ecosystems.

Early warnings about the consequences of the impacts of climate change on forests have been documented in, among others, *The Carbon Bomb: Climate change and the fate of the northern Boreal forests* – a 1994 Greenpeace report which states on page 2 that:

'Studies on the global carbon cycle suggest that boreal forests are not absorbing as much carbon as they did before 1976. As a result, the atmosphere already appears to contain 10-15 billion tonnes of carbon more than it would have if forests had continued to absorb carbon at the pre-1976 rate. If boreal forests continue to decline, estimates suggest that burning and rotting of boreal forests could contribute to the release of up to 225 billion tonnes of extra carbon into the atmosphere, increasing current levels by a third. This would accelerate the rate of climate change.'

While it is possible that the boreal forest could expand into the frozen tundra as temperatures increase, such an expansion would likely be delayed by slow tree migration rates. Even in the long-term, the boreal forest is unlikely to move northward fast enough to compensate for the breakdown of boreal forests at the southern part, turning dense forest into open woodlands and grassland, which in turn will result in a lowered biological diversity and a reduced ability of these ecosystems to store carbon and water.

Other forest ecosystems are faced with a similar fate. According to the Intergovernmental Panel on Climate Change (IPCC), a UN panel of climate scientists, it is likely that many tree species will not be able to change their geographic distribution fast enough to keep up with projected shifts in suitable climate and increased rates of extinctions are expected to occur.

What can be done?

So trees (or the lack of them) are one of the key problems that must be tackled if we are to minimise the effects of climate change. It is not surprising then that many have also suggested planting trees (carbon sinks) or reducing emissions from deforestation and degradation (REDD) are among the solutions to our climate crisis. SinksWatch believes, however, that while urgent action is needed to halt forest loss and restore degraded forests, both of these concepts – REDD and carbon sinks – are a dangerous distraction from tackling climate change. Instead of addressing the root causes of the forest as well as the climate crisis, REDD and carbon sinks are concepts that are linked to a global carbon market where planting trees or reducing deforestation in one place (usually the global South) justifies even more fossil fuel emissions somewhere else (usually in

SINKSWATCH

an industrialised country). And many of these carbon sink projects have financed the expansion of large-scale industrial tree plantations. SinksWatch believes that efforts to tackle forest loss must not justify more fossil fuel emissions because such a trade-off would neither help avert runaway climate change nor help save forests in the long run. Runaway climate change will also have a major negative impact on the world's forests.

⇨ The above information is reprinted with kind permission from SinksWatch. Visit www.sinkswatch.org and www.fern.org for more information.

© SinksWatch

Greenland ice sheet faces 'tipping point in ten years'

Scientists warn that a temperature rise of between 2°C and 7°C would cause ice to melt, resulting in a 23-ft rise in sea level.

By Suzanne Goldenberg

The entire ice mass of Greenland will disappear from the world map if temperatures rise by as little as 2°C, with severe consequences for the rest of the world, a panel of scientists told [the US] Congress today.

Greenland shed its largest chunk of ice in nearly half a century last week, and faces an even grimmer future, according to Richard Alley, a geosciences professor at Pennsylvania State University.

'Sometime in the next decade we may pass that tipping point which would put us warmer than temperatures that Greenland can survive,' Alley told a briefing in Congress, adding that a rise in the range of 2°C to 7°C would mean the obliteration of Greenland's ice sheet.

The fall-out would be felt thousands of miles away from the Arctic, unleashing a global sea-level rise of 23 feet (7 metres), Alley warned. Low-lying cities such as New Orleans would vanish.

'What is going on in the Arctic now is the biggest and fastest thing that nature has ever done,' he said.

Speaking by phone, Alley was addressing a briefing held by the House of Representatives Select Committee on Energy Independence and Global Warming.

Greenland is losing ice mass at an increasing rate, dumping more icebergs into the ocean because of warming temperatures, he said.

The stark warning was underlined by the momentous break-up of one of Greenland's largest glaciers last week, which set a 100-sq-mile chunk of ice drifting into the North Strait between Greenland and Canada.

The briefing also noted that the last six months had set new temperature records.

Robert Bindschadler, a research scientist at the University of Maryland, told the briefing: 'While we don't believe it is possible to lose an ice sheet within a decade, we do believe it is possible to reach a tipping point in a few decades in which we would lose the ice sheet in a century.'

The ice loss from the Petermann Glacier was the largest such event in nearly 50 years, although there have been regular and smaller 'calvings'.

Petermann spawned two smaller breakaways: one of 34 sq miles in 2001 and another of 10 sq miles in 2008.

The entire ice mass of Greenland will disappear from the world map if temperatures rise by as little as 2°C, with severe consequences for the rest of the world

Andreas Muenchow, Professor of Ocean Science at the University of Delaware, who has been studying the Petermann Glacier for several years, said he had been expecting such a break, although he did not anticipate its size.

He also argued that much remains unknown about the interaction between Arctic sea ice, sea level, and temperature rise.

Muenchow told the briefing that over the last seven years he had only received funding to measure ocean temperatures near the Petermann Glacier for a total of three days.

He was also reduced, because of a lack of funding, to paying his own air fare and that of his students so they could join up with a Canadian ice-breaker on a joint research project in the Arctic.

10 August 2010

© Guardian News and Media Limited 2010

THE GUARDIAN

The complicated truth about sea-level rise

Just like the swimming polar bears have become symbols for disappearing sea ice in the Arctic, the remote atolls of the Pacific and the Indian Ocean have become emblematic for the consequences of sea-level rise.

By Eilif Ursin Reed

It makes plain sense that on islands where the highest elevation is sometimes less than two metres, the IPCC's predicted sea-level rise of up to 58 cm by 2100 will cause devastation.

Or does it? Things that seem obvious at first glance, usually turn out to be more complicated if you look closer. So too with climate change.

In spite of the attention devoted to tropical islands over the last few years, with stories about 'climate refugees' and whole nations being forced to move off their islands because of sea-level rise, research on the subject has been scarce.

There is little doubt that climate change is happening. It is highly probable that we will experience sea-level rise this century and with it an increase in severe flooding events. However, how and to what extent this will affect the islanders of the world is a different story. Not necessarily a happier one, but it could at least be one about capabilities, adaptation, knowledge and resilience; rather than the one-sided doomsday narratives we have come to know too well.

New information for old assumptions

The Many Strong Voices (MSV) project is a constellation of researchers and organisations aiming to communicate the complexities of climate change, and to gather information that makes islanders capable of making knowledge-based decisions.

Earlier this year, Arthur P. Webb from MSV partner, the Pacific Islands Applied Geoscience Commission (SOPAC), and Paul S. Kench from the University of Auckland contributed to shedding new light on one of the most persistent narratives of them all, that of 'the sinking islands'. This has been a popular story among activists, journalists and politicians.

One year ago, leading up to the climate change conference in Copenhagen, Maldives president Mohammed Nasheed arranged a cabinet meeting under the rolling waves of the Indian Ocean. The aim of the publicity stunt was to communicate that sea-level rise due to climate change threatens to submerge the Maldives and other islands. Already then, Paul S. Kench told Associated Press that the outlook for the Maldives is 'not all doom and gloom'.

'The islands won't be the same, but they will still be there,' he said, pointing out that his studies of the Maldives showed that the islands can adjust their shape in response to environmental changes.

But there is still much uncertainty as to how islands react to rising sea levels. Even though considerable scientific effort has been put towards reconstructing the past and present sea-level behaviour, research on atoll island change has been scarce. While there is a good system for gathering data on sea-level trends on a global scale, there is no systematic monitoring programme to document reef island change. The researchers note that this 'seems a gross oversight given the international concern over small island stability and pressing concerns of island communities to manage island landscapes'.

Webb and Kench decided to study island change on a relatively short time scale. By meticulously examining satellite and aerial photos (shot over the last 19-61 years) of 27 atoll islands in the Central Pacific, where local sea levels have risen 120 millimetres (4.8 inches) over the past 60 years, the researchers found that 86% of the islands they selected (which didn't include the Maldives) either had grown in size or remained stable. Even severe flooding events, such as the tsunami in 2004, had added to some of the islands' size.

Encouraging the discouraged

This was welcome news to those already sceptical about climate change. *The Washington Times* said that this was yet another piece of news that exposed 'Al Gore's fairy tales', while the British magazine *Spectator* urged the Maldivian president to sell his snorkel.

The researchers involved in the study later told Germany's *Der Spiegel* magazine that they found this polarisation unfortunate, as they take global warming very seriously. They were, after all, not saying that islands are unaffected by climate change, just that in the case of sea-level rise, things are more complicated than many people seem to think. It is also important to remember that their article solely dealt with sea-level rise, which is only one of many effects of climate change, and selected islands in the Pacific.

So how does climate change affect the islands of the world? To this question there are as many answers as there are islands. There are 52 small island developing states (SIDS) recognised by the UN. These face similar

CICERO

challenges concerning freshwater supply, limited land-based resources, remoteness and vulnerability to disasters, including climate change.

Yet there are great differences between them: some have a few thousand inhabitants, such as Tuvalu, while others, such as Papua New Guinea, have millions (including the storm surge-battered Carterets). An island state like The Maldives is mainly coral islands, while Montserrat is volcanic; Tokelau is low lying while Cape Verde is mountainous; some are stable democracies while others are struggling with internal conflict. Some of the SIDS, like Belize and Guyana, aren't even islands: they are coastal states sharing many of the same characteristics and challenges as island states.

Also, you need more than just firm ground under your feet to live on an island. Though some islands are growing, the shorelines may be growing faster than the island's interior. This could leave both arable land and important freshwater supplies below sea level, rendering them susceptible to contamination by salt water.

Know the whole story

It is apparent that many circumstances need to be factored in when assessing an island's vulnerability to climate change. The consequences of climate change on small island states can be compared to those of an earthquake. An earthquake of the same magnitude and depth may cause different degrees of damage, depending on where it strikes, as we saw in Haiti and Chile. One killed hundreds, the other hundreds of thousands.

Population density, governance structures and whether the infrastructure is built to withstand earthquakes help to decide the severity of the disaster.

As with earthquakes, a lot can be done to minimise the impacts of climate change and rising sea levels. And as with earthquakes, how well prepared a community is for an event determines the severity of the outcome. Consider a situation where a flooding event damages important infrastructure. Is it fair to blame climate change if the damages could have been avoided with proper funding for adaptation?

Adaptation can be physical measures such as constructing flood barriers, moving habitation inland or to higher ground. But it can also be to provide education and training to fishermen and farmers that could help them reduce their vulnerability to climate change. Poor sanitation is another example that leads to shortages in limited freshwater supplies. Knowledge on how to manage freshwater supplies can thus reduce a community's vulnerability to the effects of climate change.

Which measures are most suitable depend on the specific challenges posed to the different islands. They also depend on what local knowledge already exists on the island in question. Consequently, finding suitable measures requires a combination of local traditional knowledge and scientific knowledge. As in Samoa, where the government spoke to and consulted local villages and people when designing and implementing coastal management plans.

In Tuvalu, on the other hand, researchers found that local inhabitants are sceptical of climate change and are not involved in discussions concerning potential adaptation options. The Tuvalu example points to a need to strengthen information efforts in order to increase local knowledge on climate change and to involve Tuvalu's inhabitants in their own future.

Understanding how climate change affects small island states is a laborious affair, that involves a plethora of voices, needs and solutions. Listening to one message is more convenient than relating to a cacophony of voices telling different stories. Yet, in the case of climate change, and how it affects the lives of people all over the world, we don't have a choice. All voices must be heard.

17 November 2010

Chapter 2

THE CLIMATE DEBATE

Unscientific hype about the flooding risks from climate change will cost us all dear

The warmists have sound financial grounds for hyping the dangers of flooding posed by climate change, writes Christopher Booker.

A s the great global warming scare continues to crumble, attention focuses on all those groups that have a huge interest in keeping it alive. Governments look on it as an excuse to raise billions of pounds in taxes. Wind farm developers make fortunes from the hidden subsidies we pay through our electricity bills. A vast academic industry receives more billions for concocting the bogus science that underpins the scare. Carbon traders hope to make billions from corrupt schemes based on buying and selling the right to emit CO_2. But no financial interest stands to make more from exaggerating the risks of climate change than the re-insurance industry, which charges retail insurers for 'catastrophe cover', paid for by all of us through our premiums.

> *The data show no evidence of an increase in UK rainfall at all. Any idea that there is one seemed to be entirely an artefact of the computer models*

An insight into this was given by a paper published by *Nature* on 17 February, which claimed to show for the first time how man-made climate change greatly increases the risk of flood damage. Among the eight authors of the paper are two of the most influential scientists at the heart of the UN's Intergovernmental Panel on Climate Change, Professor Peter Stott of the UK Met Office's Hadley Centre and Dr Myles Allen, head of Oxford's Climate Dynamics Group. Two of their co-authors are from Risk Management Solutions (RMS), a California-based firm which is the world leader in advising the insurance industry on climate change.

The study, based entirely on computer models, focused on the exceptional flooding that took place in England and Wales in the autumn of 2000. Its conclusion – that climate change could increase the chance of flooding by up to 90 per cent – was widely publicised, without questioning, by all the usual media cheerleaders for global warming, led by the BBC's Richard Black ('Climate change increases flood risk, researchers say').

When less partisan observers examined the paper, however, they were astonished. Although *Nature* has long been a leading propagandist for man-made climate change, this example seemed truly bizarre. Why had this strangely opaque study been based solely on the results of a series of computer models – mainly provided by the Hadley Centre and RMS – and not on any historical data about rainfall and river flows?

The Met Office's own records show no upward trend in UK rainfall between 1961 and 2004. Certainly autumn 2000 showed an unusual rainfall maximum, but it was

exceeded in 1930. The graph between then and 2010 shows no significant upward trend. While 2000 may have seen a lot of rain, 1768 and 1872 were even wetter. In the real world, the data show no evidence of an increase in UK rainfall at all. Any idea that there is one seemed to be entirely an artefact of the computer models.

On Friday came the fullest and most expert dissection of the *Nature* paper so far, published on the Watts Up With That website by Willis Eschenbach, a very experienced computer modeller. His findings are devastating. After detailed analysis of the study's multiple flaws, he sums up by accusing *Nature* of 'trying to pass off the end result of a long daisy-chain of specifically selected, untested, unverified, un-investigated computer models as valid, falsifiable, peer-reviewed science'.

His conclusion is worth quoting at some length: 'When your results represent the output of four computer models, fed into a fifth computer model, whose output goes to a sixth computer model, which is calibrated against a seventh computer model, and then your results are compared to a series of different results from the fifth computer model, but run with different parameters, in order to show that flood risks have increased from greenhouse gases...' you cannot pretend that this is 'a valid representation of reality', let alone 'a sufficiently accurate representation of reality to guide our future actions'.

This is precisely why the *Nature* study is of such significance – because it will undoubtedly be used to guide future actions, which will in one way or another impact on all our lives.

For a start, consider the players in this drama. Professor Stott and Dr Allen have long been among the most influential scientists in the world in stoking up climate alarmism. A famous analysis by John McClean showed that they played a key part in compiling the single most important chapter in the IPCC's last report, in 2007. The chapter entitled 'Understanding and attributing climate change' cited many more papers by them than anyone else. They have now been appointed as lead authors of the relevant chapter in the next IPCC report, *Detection and attribution of climate change*, which will guide the actions of governments all over the world.

As for their two colleagues from RMS, this is not the first time that this leading adviser to the world's re-insurance industry has been involved in a controversial bid to heighten alarm over the consequences of climate change.

In October 2005, in the wake of the Hurricane Katrina disaster, RMS held a meeting in Bermuda with four hurricane specialists, all of the alarmist persuasion, to quiz them as to how they thought hurricane activity was likely to be affected between 2006 and 2010, thanks to climate change, and how this would impact on the

southern United States, notably Florida. On the basis of this meeting, RMS advised the re-insurers that the risk of hurricane damage over the next four years was hugely increased. The companies found that their reserves were \$82 billion short of what they might be expected to pay. Premiums, particularly in Florida, accordingly rocketed upwards.

Under the heading 'The \$82 billion prediction', the details of this episode are chronicled on his blog by Dr Roger Pielke Jr, who in 2008 advised RMS that the methodology on which it relied was so biased that 'a group of monkeys would have arrived at the exact same results'. Dr Pielke, an expert in environmental impacts, recently published a chart showing how, although the RMS prediction for hurricane damage between 2006 and 2010 was a third higher than the historical average, the actual cost proved to be well under half the average figure. But, thanks to RMS, the insurance industry had made billions from higher premiums.

In 2008, following the disastrous floods of summer 2007, that vociferous climate alarmist Bob Ward, now at the Grantham Institute but then a director of RMS, called for the British Government to work more closely with the insurance industry 'to devise mutually beneficial strategies for dealing with flood risks'. We understand how working with RMS might be beneficial to the insurance industry. But whether, in light of the *Nature* study, the Government would find it beneficial is another matter – never mind the rest of us, as we are asked to pay ever higher insurance premiums, based not least on the findings of those RMS computer models.

26 February 2011

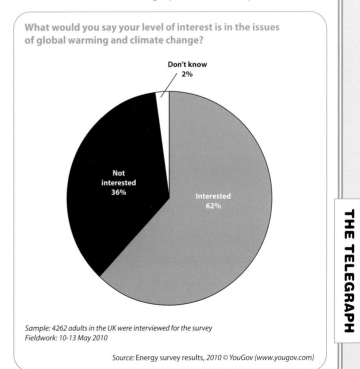

What would you say your level of interest is in the issues of global warming and climate change?

Don't know 2%

Not interested 36%

Interested 62%

Sample: 4262 adults in the UK were interviewed for the survey
Fieldwork: 10-13 May 2010

Source: Energy survey results, 2010 © YouGov (www.yougov.com)

Ten facts on climate science

Climate change is happening, and it's because of us. Here are ten facts you should know.

People say the world isn't really getting warmer, some years are just hotter than others, and it varies/goes around in cycles

The first decade of this century has been, by far, the warmest decade on the instrumental record. Despite 1998 being the warmest individual year, the last ten years have clearly been the warmest period in the 160-year record of global surface temperature.

Over the last 100 years the Earth has warmed by about 0.75 degrees Celsius and the speed it is warming at is getting faster. These days the UK Spring arrives about ten days earlier than it did in the 1970s. In 160 years of records, the ten hottest years have been in the last 13 years.

Arctic sea ice is melting; the September summer minimum extent has shrunk by about 10% every ten years since the late 70s. The smallest amounts of Arctic summer ice on record were in the last three years: 2007-2009. In a few decades, large parts of the Arctic Ocean are expected to have no late summer sea-ice at all.

People say we've nothing to do with it

Carbon dioxide (CO_2) levels in the atmosphere have gone up 38%, to 387 ppm, since pre-industrial times. Rising levels of greenhouse gases are directly linked to human activity like burning fossil fuels and clearing forests. There is a clear link between more greenhouse gases in the atmosphere and global warming.

But not all scientists agree though, right?

The overwhelming majority of climate scientists agree that human-induced climate change poses a huge threat to the world. The Intergovernmental Panel on Climate Change is not run by any government – 'intergovernmental' means it answers to all 192 governments signed up to it. Its reports are written by independent scientists and it is one of the most rigorous scientific bodies that exists. It brings together many thousands of scientists from countries all over the world to put together the best assessments of climate science available.

What about the leaked emails from the University of East Anglia? Don't they undermine the science?

There is an independent review looking at this incident.

But there is an overwhelming consensus, based on decades of climate science and the work of thousands of scientists around the world, which says that climate change is real and a major threat.

It's too late, we just need to accept it

The scientific consensus says we need to stop the world getting more than two degrees warmer than pre-industrial times if we want to avoid dangerous climate change. After that, in many regions, it will become harder to produce food, and competition for water and rises in sea level and loss of species will get much worse. We've got the technologies we need for a low-carbon world: we just need to go for it now. It'll cost much less to go low-carbon than it will to let climate change happen.

A bit of melting ice and slightly hotter summers, what's the problem?

Global sea levels have already risen by about 17 centimetres since 1900, thanks to melting ice and warming oceans. This is already threatening low-lying countries, such as islands and Bangladesh. Millions more

DEPARTMENT OF ENERGY AND CLIMATE CHANGE

people are expected to be flooded every year by 2080. The global sea level could rise by up to 59 centimetres this century. In Europe alone this could affect over 20 million people. And it looks like the sea is rising more quickly now than in the 20th century.

Some countries have always had droughts, it's nothing new

Severe droughts are now twice as common as they were in 1970. More drought is affecting which crops we can grow effectively. Global demand for food is expected to nearly double by 2050, but lack of water could mean the world produces less food, not more.

Global warming is just to do with natural changes in the Sun

Scientists are clear: there is strong evidence that changes in solar radiation could not have caused the rapid warming we have seen over the past 50 years. Since the Industrial Revolution, additional greenhouse gases have had about ten times the effect on the climate forcing as changes in the Sun's output.

We've all got a lot on our plate – let's worry about it later

Even if all greenhouse gas emissions stopped tomorrow, we are already locked into a global temperature rise of at least 1.4°C (since 1750) because of the delayed impact between emissions and temperature change. It is already happening, and we need to act now to stop it getting much worse.

It won't happen to us though

Developed countries suffer impacts too. The 2003 heatwave in western Europe, which caused 35,000 deaths (2,000 in the UK), is already twice as likely to happen again next year. By the 2040s Europe will consider such a summer normal. By the 2060s they will consider it cool.

Surely it's only the odd polar bear, who cares?

Species are already being forced to migrate or adapt. Scientists think that around 20% of species will become extinct with two degrees of warming – and it will be a real challenge, even if we act right now, to keep to that limit.

⇨ The above information is reprinted with kind permission from the Department of Energy and Climate Change. Visit www.decc.gov.uk for more information.

© Crown copyright

Climate change blame?

Summary of a YouGov survey.

The majority of the British public believe that Planet Earth is experiencing climate change, but few place the blame entirely on human activity. A significant 84% agreed with the statement that the planet is warming, but only 18% believe human activity is mainly responsible; most (58%) feel that other factors have a part to play. 8% think that human activity, in comparison to other factors, is not responsible at all.

A small but noticeable 8% refute the idea that the planet is warming at all.

It seems that most Britons also believe that politicians are not doing enough against global warming with 73% agreeing with the statement that 'there is no serious political will worldwide to tackle climate change' – perhaps due to the disappointing results of the Copenhagen summit which had been hailed firstly as the great 'climate change' conference, and then as a major disappointment due to its failure to reach consensus between the participating nations.

It seems that the recent controversy surrounding scientific evidence supporting the extent, and reasons behind, the warming, dubbed 'Climategate' by the BBC, has created a huge amount of uncertainty around the issue. In fact, 30% of the public claim that they have become more sceptical about the validity of climate change and 38% agree that dishonest scientists have made them doubt whether climate change is really happening.

9 April 2010

⇨ The above information is reprinted with kind permission from YouGov. Visit www.yougov.com for more information.

© YouGov

DEPARTMENT OF ENERGY AND CLIMATE CHANGE / YOUGOV

Are we still sure about climate change?

Climate scientists have had a poor press in recent months. Stuart Parkinson investigates whether this is a sign that the scientific evidence for climate change is less robust, or just media misrepresentation.

Over the last year or so, climate science has been heavily criticised in the media, especially in the UK and USA. Particular criticism has been directed at researchers at the Climate Research Unit at the University of East Anglia (UEA) and also at the Intergovernmental Panel on Climate Change (IPCC), which summarises the evidence of climate change, its causes and its potential effects, for international policy-makers. So, has the criticism been justified? Is the scientific evidence on the threat of climate change less robust than previously claimed?

Stolen emails and 'hidden' data

The first wave of criticisms surfaced in November 2009 in the run-up to the Copenhagen climate negotiations, when about a thousand private emails were stolen from a server at UEA and released online. These emails included correspondence between some leading climate scientists over the previous 13 years, including the Director of UEA's Climate Research Unit, Professor Phil Jones, and Professor Michael Mann of Pennsylvania State University.

Of the emails released, a small minority contained comments that were used to question the integrity of the scientists involved. For example, one of most widely circulated emails (written back in 1999) included the comment, 'I've just completed Mike's *Nature* [the science journal] trick of adding in the real temps to each series for the last 20 years (i.e. from 1981 onwards) and from 1961 for Keith's to hide the decline'. Using a 'trick' to 'hide the decline' was interpreted by climate sceptics as evidence that data had been massaged to hide a true decline in global temperatures.

However, the explanation for these comments was far more mundane. The 'trick' referred to was simply shorthand for 'an effective methodology for processing the data', and the 'decline' being 'hidden' was a well-known (at least within climate science circles) problem with a particular tree-ring data-set, which diverged from other comparable temperature data-sets. Other 'suspect' comments within the emails were similarly innocuous, as a number of climate scientists pointed out, although a few caused raised eyebrows due to their belligerent tone.

Nevertheless, the media furore caused by these emails – dubbed 'climategate' – was such that four separate investigations were carried out into the concerns during 2010. Three were carried out in the UK – one by the House of Commons Science and Technology Committee, and two commissioned by the UEA but carried out independently. These focused mainly on the conduct of researchers at the Climate Research Unit, with the third review in particular going into considerable detail. The fourth, meanwhile, was carried out in the USA and focused on Michael Mann's research.

> *Reports in the mainstream media in the UK over the past year or so have given the distinct impression that evidence for the threat of climate change is less than clear*

The reviews rejected allegations that climate scientists had colluded to withhold scientific evidence, interfered with the peer-review process to prevent dissenting scientific papers being published, deleted raw data, or manipulated data to make the case for climate change appear stronger than it is. There was, however, some limited criticism, especially regarding inadequate responses to data requests under the Freedom of Information Act.

Glaciers and the IPCC

With the debate about climategate still reverberating around the web, climate scientists were hit by another allegation in mid-January 2010. This one was directed at the IPCC and, in particular, a claim in its landmark 2007 report that Himalayan glaciers could melt away by 2035. The IPCC quickly admitted that this was a mistake that had crept into a paragraph in volume two of the report, but argued that its overall conclusions concerning the problems of melting glaciers in the 'summary for policymakers' remained valid. Indeed, volume one of the report had been accurate in its reporting of the research on glacial retreat in the Himalayas. As journalists questioned whether the mistake undermined the IPCC's credibility, vice-chair of the IPCC, Jean-Pascal van Ypersele, was quoted as saying, 'I don't see how one mistake in a 3,000-page report can damage the credibility of the overall report.'

However, this was not enough to quell critics. Not only was this mistake used by numerous commentators to question the validity of the whole report, further allegations of IPCC 'mistakes' and improper conduct of climate scientists were made, especially in *The Sunday Times*, but also in some other British newspapers. The alleged mistakes concerned issues such as the threat to crop yields in Africa due to climate change, possible links between trends in natural disasters and climate change, and the vulnerability of the Amazon rainforest. As climate scientists pointed out – for example on the RealClimate website – there was generally little substance to any of the criticisms.

Let us take as an example the criticism of the IPCC claim that 'up to 40% of the Amazonian forests could react drastically to even a slight reduction in precipitation'. In an article in *The Sunday Times* at the end of January, journalist Jonathan Leake alleged the claim was 'bogus', arguing that the IPCC had misrepresented research, and quoted British climate researcher, Simon Lewis, to back up his allegations. Unfortunately for Leake, Lewis filed a complaint stating that his views had been misrepresented in the article. *The Sunday Times* upheld the complaint, acknowledged that the article had incorrectly criticised the IPCC and removed the piece from its website. Unfortunately, the retraction took place over four months later, so the damage to the credibility of climate science had been done.

To deal with the rising concerns about the accuracy of IPCC reports, the UN Secretary-General requested the Inter-Academy Council (IAC), an umbrella group of many of the world's most prestigious science academies, to carry out a review of the IPCC's internal procedures. The IAC reported in August, concluding that 'the IPCC assessment process has been successful overall'. However, it did make a series of recommendations to improve the robustness of future reports, including adopting clear procedures on conflicts of interest.

Recent climate science

Amongst this frenzy of debate over the integrity of climate scientists, media reporting of climate science itself has taken a back seat. Nevertheless, the evidence continues to mount about the extent of the threat.

For example, papers in a special issue of a Royal Society journal published online in November 2010 examine how quickly the world may reach 4°C of warming above the pre-industrial average, as well as the impacts this may bring. They conclude that a 'business as usual' scenario could yield such a change as early as the 2060s, with considerable impacts on, for example, water availability and crop yields.

If there is so little substance behind the media criticisms of the last year, one must ask how such stories became so prominent

Meanwhile, although the UK has experienced unusually cold temperatures during recent winters, the globally averaged temperature continues to be exceptionally high, with 2010 set to be among the three highest years on record. Moreover, the decadal average of global temperature – a more reliable indicator than annual average – has been markedly higher in the first decade of the 2000s than in any previous decade on record.

Another study that examined 'expertise' in climate science is also notable. It analysed the views and publication records of over 1,300 climate researchers. Of those with the highest rate of climate science publications, 97-98% were convinced that human activities were causing climate change. Sceptical researchers had much lower levels of expertise in the field.

Sceptics and public opinion

If there is so little substance behind the media criticisms of the last year, one must ask how such stories became so prominent. A detailed examination is beyond the scope of this article, but one facet has certainly been the influence of leading free-market advocates and their allied think tanks, which oppose new regulations enacted in the name of climate change. It is notable, for example, that Richard North – fellow of the Institute for Economic Affairs – carried out the research for *The Sunday Times* article on the Amazon discussed above. Meanwhile, former Conservative Chancellor Nigel Lawson – who founded the climate sceptic organisation the Global Warming Policy Foundation – is frequently invited to give his views in the media. That is not to say that all climate sceptics are free-market champions, but without this very powerful lobby it is unlikely such a media storm would have been created. SGR has also highlighted the powerful role of the oil industry in supporting free-market think tanks in their promotion of climate sceptic ideas over the last two decades.

How much has this corrosive media coverage undermined public belief in climate change, and support for action to tackle it? Results of recent opinion polls provide some interesting answers. One conducted in early 2009 by the University of Cardiff and Ipsos MORI showed that, while public concern about climate change has fallen, it was nevertheless still high – over 70%. In addition, only 20% believed there was serious disagreement among scientists over whether climate change is caused by humans. Another poll, commissioned by BBC News at a similar time, showed that the increase in doubt over global warming was due to the cold winter and not the scientific controversies.

Conclusions

Reports in the mainstream media in the UK over the past year or so have given the distinct impression

that evidence for the threat of climate change is less than clear. But an investigation of the facts behind the headlines, coupled with an examination of the academic research, reveals that this is anything but the case. While significant uncertainties in the science do exist, the defining aspects of the problem – that climate change is happening, that it is mainly caused by human activities, and that it is likely to have very serious impacts if left unchecked – remain solidly backed by the data.

Nevertheless, it is clear that attempts to discredit the science of climate change will continue. Although sceptics have had limited success so far, with more unusually cold weather this winter in the UK, their hand will be strengthened. Hence, organisations like Scientists for Global Responsibility need to continue to challenge unbalanced media coverage. Meanwhile, although the basic evidence is robust, climate scientists do need to deal with some of the weaknesses in their research activities – especially concerning openness with data. These actions will allow us to overcome misinformation and thus keep up the pressure on policy-makers to take the necessary action to bring about a rapid reduction in greenhouse gas emissions.

Dr Stuart Parkinson is Executive Director of SGR. He holds a PhD in climate science, and was an expert reviewer for the IPCC from 1999-2001.

Winter 2011

⇨ The above information is reprinted with kind permission from Scientists for Global Responsibility. Visit www.sgr.org.uk for more information.

© *Scientists for Global Responsibility*

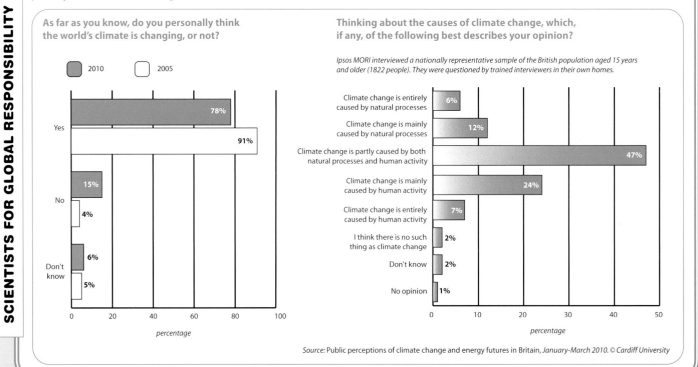

SCIENTISTS FOR GLOBAL RESPONSIBILITY

As far as you know, do you personally think the world's climate is changing, or not?

2010   2005

Yes — 78% (2010), 91% (2005)
No — 15% (2010), 4% (2005)
Don't know — 6% (2010), 5% (2005)

percentage

Thinking about the causes of climate change, which, if any, of the following best describes your opinion?

Ipsos MORI interviewed a nationally representative sample of the British population aged 15 years and older (1822 people). They were questioned by trained interviewers in their own homes.

Climate change is entirely caused by natural processes — 6%
Climate change is mainly caused by natural processes — 12%
Climate change is partly caused by both natural processes and human activity — 47%
Climate change is mainly caused by human activity — 24%
Climate change is entirely caused by human activity — 7%
I think there is no such thing as climate change — 2%
Don't know — 2%
No opinion — 1%

percentage

Source: Public perceptions of climate change and energy futures in Britain, January-March 2010. © Cardiff University

Climate change scepticism

A summary of a survey carried out by YouGov.

By Zara Atkinson

The British public is becoming less, not more, concerned about climate change, a survey for energy company EDF has found. Considering the structure of the new Conservative-Liberal Democrat Coalition Government, it is interesting to note a significant difference in the views of the two parties' supporters on these issues. Liberal Democrat supporters are the most likely to be interested in global warming and climate change, at 79%, compared to 53% of Conservative supporters.

One-third (33%) of the public now agrees with the statement 'it is not yet clear whether climate change is happening or not'

The survey results indicate that concern for global warming and climate change is continuing to decrease. Although 28% of the public agreed with the statement that 'it is a serious and urgent problem and radical steps must be taken NOW to prevent terrible damage being done to the planet', this is down from 38% who agreed in 2007; a 10% drop in three years. Again, far more Liberal Democrat supporters think this way (43%) than supporters of the Conservatives (18%).

'Not clear whether climate change is happening or not'

Given the intense media attention surrounding leaked documents from the University of East Anglia's Climatic Research Unit, which suggested that data that was not consistent with theories supporting global warming were being deliberately withheld, it is perhaps not surprising that more are becoming sceptical about the existence of climate change. One-third (33%) of the public now agrees with the statement 'it is not yet clear whether climate change is happening or not – scientists are divided on this issue', compared to only 25% in 2007. There also seem to be many more sceptics among Conservative supporters, 43% of whom agreed with the statement. Only 20% of Liberal Democrat supporters thought the same.

Nuclear power, hailed by some as a green power alternative, appears to be gaining support as a source for Britain's future energy. As most of Britain's older nuclear and coal power stations will come to the end of their useful lives by about 2020, a substantial 'power gap' will need to be filled. 52% of Britons now support the gap being filled by new nuclear power stations (up 6% from 46% in 2007), while 64% agree with the statement 'nuclear energy has disadvantages but the country needs it as part of the energy balance, with coal, gas and wind power' (an increase of 5% from 59% in 2007). Just 18% now agree that 'the most important thing is stop building nuclear power stations' compared to 24% in 2007. And while the source of power may be in dispute, the need for Britain to be self-sufficient in energy seems irrefutable: 90% now agree with this, exactly the same as did in 2007.

Cross-party support

And it seems that nuclear power is an attractive option across all parties as well. Despite Lib Dem supporters being slightly less in favour of nuclear power than their Labour or Conservative counterparts, they still nevertheless agree with all the statements suggesting the use of nuclear power. The majority of Liberal Democrat supporters agree that nuclear energy is still 'needed as part of the energy balance with coal, gas and wind power' (58%, compared to 64% for Labour and 74% for the Conservatives), and that old nuclear power stations should be replaced by new ones and other renewable energy sources (60% compared to 66% and 67%). There is also net agreement (i.e. the percentage that agree subtracted from the percentage that disagree) amongst Lib Dems that 'regulations should be changed to make it easier to build nuclear power stations' (17%). There is a net score of 26% for Lib Dem supporters disagreeing that 'the most important thing is to stop building of nuclear power stations', thereby concurring with Labour (29% net) and Conservative (50% net) supporters on this issue.

This net support goes a long way to suggest that despite decreasing levels of concern for climate change, other energy sources are being considered. It seems nuclear energy, along with an expansion of renewable sources, is seen as the most viable way forward for the UK's energy supply, and may even point to a general welcoming of the Coalition Government's recent energy proposals.

27 May 2010

⇨ The above information is reprinted with kind permission from YouGov. Visit www.yougov.com for more information.

YOUGOV

Can we really measure the climate?

Information from The Scientific Alliance.

Average temperatures or temperature ranges are often used as a simple proxy for climate. In combination with some description of rainfall, they encapsulate the essentials: in the Mediterranean it is typically hot and dry in summer and cooler and wetter in winter, and a continental climate is hot and dry in summer and cold with snow in winter, for example. But quantifying climate more precisely is fraught with difficulty.

Records kept over the years give us historical figures to make comparisons between average temperatures then and now. This sounds simple, but the very concept of an average temperature has no simple definition. First, we have to realise that temperature is what is known as an intensive property of matter. This simply means that it does not depend on the nature or size of the material for which it is measured.

Depending on the weather conditions or time of year, either the maximum or minimum temperature might be more typical of the day as a whole, yet both are implicitly given equal weight

So, for example, air and a body of water may have the same measured temperature at a particular moment, but their behaviour is very different. Air has a low thermal capacity (it take little heat to change its temperature), while water has a high thermal capacity and its temperature changes relatively slowly. In the present long cold spell in western Europe, ponds and lakes need a period of consistently sub-zero temperatures before they begin to freeze. Equally, as air temperatures rise, the ice may take many days to melt. A given volume of water has a very different thermal energy content than the same volume of air. This can be easily quantified and, in contrast to temperature, is an extensive property.

When trying to average temperatures, the first obvious rule is that the measurements must all be of the same material: you cannot average air and water temperatures, for example, and get a meaningful answer. This in itself is pretty obvious and, in discussing climate change, air and water temperatures are considered separately. However, the difficulties with averaging do not stop there.

Even if temperatures are measured under carefully controlled conditions as expected for official records, they will fluctuate quite rapidly depending on wind direction and strength, cloud cover, time of day, etc. The convention is to measure a maximum and minimum shade temperature each day. These readings can then be used to provide average maxima and minima per month or year, or combined to give an overall 'average temperature'. And the figures for individual stations can themselves be combined to give national, regional and global averages.

These figures tell us something, of course, but the desire to quantify also obscures the detail. Say, for example, that place X has an average maximum temperature of +15°C and an average minimum of +5° and place Y registers +25° and -5°. Both have an overall average of +10°, but the actual climate experienced would be quite different. In a similar way, measured air temperatures in the shade bear little relationship to the apparent temperature in the Sun. Although the measured shade air temperature might be the same whether or not the Sun is shining, the effect on the Earth's surface of the sunlight is significant and, once the ground has been warmed, it will release its heat at night to keep the air somewhat warmer, at least temporarily.

Simple averaging can be deceptive in other ways as well. Depending on the weather conditions or time of year, either the maximum or minimum temperature might be more typical of the day as a whole, yet both are implicitly given equal weight. Nevertheless, it is arguable that such issues are not important when comparing time series of measured temperatures. For example, the Central England Temperature record (CET) is the longest continual record available, with monthly means being recorded from 1659 and daily means logged from 1722. Looking at this it is easy to see the recorded range and note that temperatures do indeed appear to have been higher in the latter part of the 20th century, although they have dipped again since 2000. It is the changes which are significant rather than the absolute values, provided that all measurements are strictly comparable.

The concept of global average temperature is a little misleading, since the summary of the IPCC Fourth Assessment Report shows that the warming pattern is regional rather than global

This, of course, introduces yet another concern. The same instruments would not have been used in the 17th century as 300 years later and, with the best will in the world, it is difficult to guarantee that no artefacts have been introduced. Equally, it is hardly conceivable that the surroundings of the measuring stations will be unchanged over this period (although hopefully none of the weather stations is now in an urban area, on tarmac or near heat sources as some have been found to be in other countries).

A final problem to bring up with averages is that, to avoid giving a misleading picture, data should be taken from stations spread evenly over the Earth's surface. This is certainly not the case. In particular, there are large areas of the Arctic and Antarctic with no data being collected. The same is true for the open oceans, where collecting surface water temperatures reliably is enough of a challenge, without trying to measure air temperatures.

What we are left with then is an incomplete record of imperfect data, from which conclusions about climate change are drawn. This is the basis of the 'global warming' message. But actually the concept of global average temperature is again a little misleading, since the summary of the IPCC Fourth Assessment Report shows that the warming pattern is regional rather than global. Warming over the 20th century was recorded on all continents apart from Antarctica, but was considerably greater in the northern than the southern hemisphere. Given the greater proportion of ocean in the south, this is not surprising.

But global averages are still the main measure and this is the time of year when preliminary conclusions are drawn about the current year, as the annual meeting of the UN Convention on Climate Change takes place. So far, the message being put out by the World Meteorological Organization is that 2010 is likely to be among the warmest three on record. Based on the temperature record, this is doubtless correct, but how meaningful is this?

The WMO points towards record high temperatures in Russia, China and Greenland to support its case. Meanwhile, anyone mentioning record lows and pointing out that new records are set nearly every day somewhere in the world is told that this means nothing. In practical terms, life has to go on and adapt to whatever climatic conditions turn out to be. Measuring temperatures remains a useful thing to do, but we must be careful not to read too much into the average figures. And we should never forget that, whatever the temperature is, we still have only a hazy idea about what controls it.

⇨ The above information is reprinted with kind permission from The Scientific Alliance. Visit www.scientific-alliance.org for more information.

© The Scientific Alliance

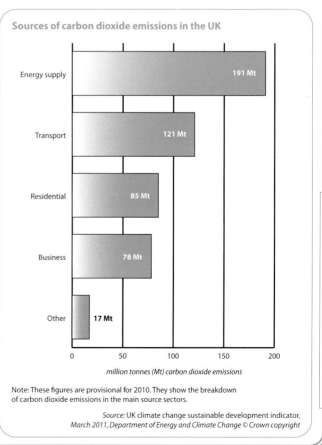

Sources of carbon dioxide emissions in the UK

- Energy supply: 191 Mt
- Transport: 121 Mt
- Residential: 85 Mt
- Business: 78 Mt
- Other: 17 Mt

million tonnes (Mt) carbon dioxide emissions

Note: These figures are provisional for 2010. They show the breakdown of carbon dioxide emissions in the main source sectors.

Source: UK climate change sustainable development indicator, March 2011, Department of Energy and Climate Change © Crown copyright

THE SCIENTIFIC ALLIANCE

Adapting to the greenhouse

Information from Panos London.

The response – what can be done?

Climate change is a global issue and needs a global response – whether creating mitigation and adaptation strategies, providing funding for these strategies, or through additional policy assistance for people who are especially vulnerable to climate change.

Mitigation and adaptation

Mitigation needs international agreement and national enforcement; emissions from one country threaten everybody, and cutting them will take time. Adaptation, on the other hand, is immediate and local: people who are affected by climate change make the changes, pay the costs and feel the benefits.

A global response to climate change needs to include both mitigation and adaptation, and have input from many quarters. For example, researchers could work on enhancing existing adaptation techniques or finding new ways to adapt (other than drastic options such as migration); journalists could then inform people of the existing options and costs, while challenging governments to build national capacity. Communities must also unite to make their voices stronger at international policy forums.

A global response to climate change needs to include both mitigation and adaptation

Countries and communities must find new ways to manage their resources – for example, by planting climate-suitable crops that tolerate or resist drought, salt and pests. Governments also need to improve planning regulations and bring in fresh policies to encourage low-energy buildings and transport systems and the use of recycled water.

Existing communication networks – both formal and informal – could play a vital role in this process. Governments can improve disaster preparedness by providing early warning of epidemics and natural disasters. The media could play a vital part in disseminating information rapidly and to a large number of people. It is also important to embrace indigenous knowledge systems, and for communities to share knowledge between themselves (for example, farmer to farmer). A Practical Action project in Zimbabwe found that radio was the most efficient mode of communication.

Reducing vulnerability

Poor people are highly vulnerable because they have little hope of earning a decent living or finding a safe place to live. Women in subsistence farming communities are particularly vulnerable. Their lack of access to, and rights over, resources and wealth such as agricultural land, combined with fragile property rights and security of tenure, leave them more exposed in a changing climate. They are often more vulnerable than men to weather-related disasters, partly because they bear a greater burden of care for their families.

Shifting policy away from reactive disaster management to more proactive capacity building can reduce gender inequality. Higher incomes, better education and technical skills, better food distribution, disaster preparedness and management and improved healthcare can all substantially reduce climate vulnerability.

Adaptation is sustainable when it is linked to effective governance, civil and political rights, and literacy – in other words, when it is part of mainstream national development planning. Insufficient information and knowledge about climate change will continue to hinder effective adaptation. More relevant information is therefore vital – about increased crop yields linked to changes in planting dates, or the cost of coastal protection investment, for example. But Asia and Africa face other urgent priorities, such as domestic conflicts, pervasive poverty, hunger, epidemics and terrorism. In these situations, it is easy for people to forget about climate change and the need to adapt.

⇨ The above information is reprinted with kind permission from Panos London. Visit www.panos.org.uk for more information.

© Panos London

PANOS LONDON

We have to adapt culturally to climate change

An article by Dr Jenny Pickerill, Senior Lecturer in Human Geography, University of Leicester.

While the case that climate change is happening and probably irreversible is robust, the political arguments about whether we should do anything about it remain ongoing and unresolved. Many in the minority world (like Britain and the USA) are relying upon technological innovation (like wind power, electric cars and geo-engineering) to save us. Research shows that while many individuals have engaged in making small changes in their lifestyle – such as using recycling bins or cycling a little more – most struggle to make big behavioural changes and feel powerless to take on the changes necessary to really mitigate further carbon emissions. Time is running out to make changes at a big enough scale to mitigate further climatic changes.

Adapting is sometimes touted as the easier option. Rather than fundamentally changing everything about how we live today, we could plan for a different future and increase our resilience to any changes ahead. There is, of course, just as much debate as to how we should prepare for these changes as there is about how to mitigate climate change. We should be careful not to be naive about what this actually involves.

For a start, climate change is likely to lead to more extreme climatic events; an unpredictability that is difficult to plan for. Predicting what climatic changes will occur is a difficult science, let alone interpreting what that means for how we live. Moreover, the magnitude of climate changes depends on whether we can reduce carbon emissions now, keep them stable, or increase them yet further. So if we are to take the adaptation route we have to plan for a variety of different scenarios and we should not abandon all attempts at mitigation. Mitigation has effectively become about trying to reduce the extent to which we need to adapt: it is not an either/or situation.

If we take housing as an example, we can begin to understand the complexity and possibility of making these changes. Housing – both in construction and use – consumes significant amounts of energy and contributes at least 25% of all carbon emissions in Britain. The Met Office predict that temperatures will rise in Britain, with increasing heatwaves and fewer frost days. At the same time we will have increased rainfall, more intensive rain showers, sea-level rise, coastal surge events and more storms. In other words, we need to be prepared for flooding, storms and heat. If we don't, then not only will our houses suffer from damage but we will continue to increase our use of energy as more people need air conditioning to keep their houses cool – creating a vicious circle of increased emissions and then greater temperature rises.

We already have the technical knowhow, and many working examples, to build resilient eco-houses in Britain. While there is a huge variety of different types of eco-homes (from those using only local natural materials to the more technological), most have been built to be resilient in today's climate, with low energy use, or autonomous (generating their own energy, collecting rainwater, dealing with their own waste, etc.).

However, there are several problems with our attempts thus far; cost, suitability for the future, retrofitting, and, most importantly, cultural understandings of the home. There remains a perception that building an eco-house is more costly, whereas figures for the lifecycle costs of buildings have proved that in the long term they are actually cheaper; we are too used to considering cost only at the build stage. But we do need to find better ways of making land, traditionally very expensive in Britain, more available for eco-building.

Index ranking of the ten largest carbon dioxide emitters

Country	Share of global CO$_2$ emissions	CCPI rank 2011	CCPI rank 2010
United Kingdom	1.74%	8	6
Korea, Rep.	1.71%	34	41
Russia	5.42%	48	45
USA	19.05%	54	53
Canada	1.88%	57	59

Note: The Climate Change Performance Index (CCPI) evaluates the climate protection performance of 57 different countries responsible for more than 90% of global carbon dioxide emissions. The lower the ranking, the lower the country's climate protection performance. Visit http://www.germanwatch.org/klima/ccpi.htm for more information.

Source: The climate change performance index, 2011 © Germanwatch

DR JENNY PICKERILL

We are building eco-housing that is suitable for today's climate and reduces carbon emissions (mitigating climate change), both of which are important, but it is not enough. We need to be designing houses which will be suitable for the future climate of wet, hot, unpredictable weather. Dr Jago Cooper, from the University of Leicester's School of Archaeology and Ancient History, has used an archaeological investigation of how pre-Columbian residents in Cuba built their houses to help inform how we might build our houses today.

Crucially, in his case study area there is evidence of abrupt climatic changes (such as flooding) which people dealt with by innovative settlement locations and building stilted wooden houses. These buildings were in the main quite flimsy and temporary lightweight structures built using easily available local materials. But they had substantive structural posts which were resilient to hurricane winds, so when storms happened most of the house was destroyed, but the main structure survived and was easily rebuilt. They had also often collectively stored food in more secure areas.

We need to be designing houses which will be suitable for the future climate

The lesson here for the modern day is that we should look beyond simply being resilient to climatic events to how we are prepared to recover and carry on afterwards. In practical terms this raises questions about whether we should be designing our houses to be more temporary or more durable, training more of us to be able to build our own houses and use more easily available local materials (like we did in the past before the development of bricks). It is also about all of us understanding the subtle balance between the need for insulation and ventilation. We need insulation to reduce draughts and keep us warm but we need ventilation to keep us cool. As the climate changes we are likely to need more ventilation than insulation, which could dramatically change the design of our houses.

As soon as we start to talk of building better houses the issue of our existing housing stock is raised. Of course we need to improve these too, but thus far we have focused on quite small changes (such as extra insulation) or adding technology to houses. We need to think more radically about how to adapt these houses to survive climate change, not just reduce carbon emissions.

Finally, adapting housing for climate change involves considerably more than technical changes to construction; it involves huge cultural shifts in how we consider our house and home. For many, a house is foremost about security – both the physical act of having somewhere safe to live and sleep, and financially

as an investment – and comfort. There is a deeply felt sense that our homes are our refuge. To change this, by making housing more temporary, using natural materials (which might be perceived as less robust), or relying on manual heating and ventilation systems, requires social changes in how we live.

Moreover, it requires us to build ready for changes that many of us have only vaguely understood to be happening; to change behaviour for an unknown future. It is not technology, or really even politics, which is holding us back in making these changes, it is deep-rooted cultural and social understandings of how we live and what we expect houses to do for us.

As a result, we can understand adapting to climate change is necessary but complex. Even just changing our housing is difficult but entirely possible. To do so we must realise that we need cultural change as much as technological and political change, and that we must ensure that those less well off have just as much opportunity to prepare as the wealthy.

⇨ The above information is reprinted with kind permission from Dr Jenny Pickerill of the University of Leicester. Visit www.leicesterexchanges.com for more.

© Dr Jenny Pickerill

Saving the world from the human response to climate change

The worst impact of climate change may be how humanity reacts to it.

The way that humanity reacts to climate change may do more damage to many areas of the planet than climate change itself unless we plan properly, an important new study published by Conservation International's Will Turner and a group of other leading scientists has concluded.

The paper *Climate change: helping nature survive the human response*, which was published today in the scientific journal *Conservation Letters*, looks at efforts to both reduce emissions of greenhouse gases and potential action that could be taken by people to adapt to a changed climate, and assesses the potential impact that these could have on global ecosystems.

In particular it notes that one-fifth of the world's remaining tropical forests lie within 50 km of human populations that could be inundated if sea levels rise by 1 m. These forests would make attractive sources of fuel-wood, building materials, food and other key resources and would be likely to attract a population forced to migrate by rising sea levels. About half of all Alliance for Zero Extinction sites – which contain the last surviving members of certain species – are also in these zones.

Dr Turner said: 'There are numerous studies looking at the impacts of climate change on biodiversity, but very little time has been taken to consider what our responses to climate change might do to the planet.'

The paper notes that efforts to reduce greenhouse gas emissions by constructing dams for hydropower generation can cause substantial damage to key freshwater ecosystems as well as to the flora and fauna in the flooded valleys. It also notes that the generally bogus concept that biofuels reduce carbon emissions is still being used as a justification for the felling of large swathes of biodiverse tropical forests.

The report also reviews studies examining the complex series of outcomes in historical examples of climate change and environmental degradation, and humanity's efforts to adapt to changing circumstances. Migration caused in part by climatic instability in Burkina Faso in the late 20th century, for example, led to a 13 per cent decline in forest cover as areas were cleared for agriculture, and a decline in fish supplies in Ghana may have led to a significant increase in bushmeat hunting.

Dr Turner added: 'If we don't take a look at the whole picture, but instead choose to look only at small parts of it, we stand to make poor decisions about how to respond that could do more damage than climate change itself to the planet's biodiversity and the ecosystem services that help to keep us all alive.

'While the tsunami in 2004 was not a climate event, many of the responses that it stimulated are comparable with how people will react to extreme weather events – and the damage that the response to the tsunami did to many of Aceh province's important ecosystems as a result of extraction of timber and other building materials, and poor choices of locations for building, should be a lesson to us all.'

Although the challenge of sustaining biodiversity in the face of climate change seems daunting, the paper notes that we must – and can – rise to the challenge. Turner adds: 'Climate change mitigation and adaptation are essential. We have to ensure that these responses do not compromise the biodiversity and ecosystem services upon which societies ultimately depend. We have to reduce emissions, we have to ensure the stability of food supplies jeopardised by climate change, we have to help people survive severe weather events – but we must plan these things so that we don't destroy life-sustaining forests, wetlands and oceans in the process.'

The paper concludes that there are many ways of ensuring that the human response to climate change delivers the best possible outcomes for both society and the environments, and notes that in particular, maintaining and restoring natural habitats are among the cheapest, safest and easiest solutions at our disposal to reduce greenhouse-gas emissions and help people adapt to unavoidable changes.

Dr Turner said: 'Providing a positive environmental outcome is often the best way to ensure the best outcome for people. If we are sensible, we can help people and nature together cope with climate change, if we are not it will cause suffering for people and serious problems for the environment.'

6 August 2010

⇨ The above information is reprinted with kind permission from Conservation International. Visit www.conservation.org for more information.

© *Conservation International*

CONSERVATION INTERNATIONAL

Technology, not targets, will save the planet

Information from the Institute for Public Policy Research (IPPR).

By Andrew Pendleton, IPPR

If climate change is the most critical issue facing mankind, it surely makes sense that world leaders come together to hammer out a binding international treaty to tackle it. This was the logic that brought Barack Obama, Gordon Brown and more than 100 other presidents and prime ministers, as well as kings, to Copenhagen this time last year.

Yet they left having put their name to nothing more substantial than a hastily drafted, face-saving political accord.

One year on, the follow-up in Cancun, Mexico, has attracted nothing like the same cast list. It has gone largely unnoticed as a result and we've crept only a fraction closer to getting agreement on an effective, co-ordinated global approach to tackling climate change. The time has surely come for a rethink.

> *If climate change is the most critical issue facing mankind, it surely makes sense that world leaders come together to hammer out a binding international treaty to tackle it*

The nub of the issue lies in the bitter winter that Chris Huhne, the energy and climate secretary, left behind when he flew to sunny Cancun. Environmentalists dismiss as crass the juxtaposition of a blast of cold weather and global climate talks – and, in truth, 2010 is on course to be one of the warmest years on record.

But it is precisely the tension between the long-term warming of the planet and the snow we see outside our windows that is undermining efforts to fashion a new treaty.

In recent weeks, faced with a choice between shivering to save mankind and turning up the heating to keep warm, most people, even those who fancy they are quite committed greens, will have chosen the latter. This is the central problem for our political leaders: when we make political choices, we prioritise the issues that immediately affect us. Global warming is a big issue, but it is not a 'now' issue for most of us.

It is not that most people are in climate change denial. Far from it. For instance, an Ipsos MORI poll earlier this year found that 78 per cent of people in Britain believe that climate change is either partly or mostly caused by human beings. Even in the more sceptical United States, polls show a majority have accepted what is now a broad consensus among scientists.

But when it comes to the crunch, other, more immediately pressing issues come first. Again, according to Ipsos MORI, at the time of the general election environmental issues were a priority for about five per cent of voters while the economy was top of the agenda for more than 70 per cent.

It's the same elsewhere. In the recent Australian election, climate change was typically rated eighth or lower in the list of voters' priorities. A US poll from early 2010 ranked climate change 21st out of 21 'priority' issues for the year ahead.

The public's attitude may be frustrating for environmentalists, but it is easy enough to see how an apparent paradox is far from irrational. Safeguarding tomorrow's climate requires making sacrifices today.

For example, a report last week from the UK's climate change committee suggested that the annual cost of decarbonising electricity supply could reach £10 billion. This cost will be passed on by energy companies to households and industry. As the annual household electricity bills in the UK already amount to about £10 billion and count for about one-third of total electricity use, the maths is pretty simple.

This is why global climate talks make at best only glacial progress. World leaders cannot go and make grand gestures because they have to answer to their public back home. And they simply do not have the consent for action. A new approach is urgently needed.

Climate change shroud-waving isn't working. Being warned of the dire consequences of inaction has not proved a game-changer. So the first task is to look more intently at what might win public support.

A poll conducted by IPPR before the May election found that, while voters were less willing to accept changes in the way energy was produced on environmental grounds, they thought our supply needed to be more secure and were in favour of using clean technology such as solar and wind power.

More fundamentally, the economy is always high on most people's agenda and so it is critical that climate-friendly policies work to achieve prosperous – and growing – economies, not to cut across them.

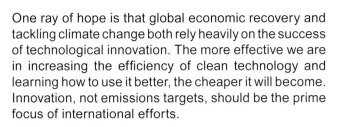

One ray of hope is that global economic recovery and tackling climate change both rely heavily on the success of technological innovation. The more effective we are in increasing the efficiency of clean technology and learning how to use it better, the cheaper it will become. Innovation, not emissions targets, should be the prime focus of international efforts.

It's time to turn away from these annual environmental shindigs and concentrate on territory about which people care most. The task is no longer about targets to cut emissions, but about defining how a more secure and growing economy can also be a low-carbon economy.

12 December 2010

⇨ The above information is reprinted with kind permission from the Institute for Public Policy Research (IPPR). Visit www.globalclimatenetwork.info for more information.

Climate change victory for Huhne

Britain has committed to the most comprehensive climate change plan in the world by setting a 'carbon budget' to halve greenhouse gases by 2025.

Chris Huhne, the Energy and Climate Change Minister, said the UK would drastically speed up green measures from 2020, ultimately cutting carbon by 60 per cent by 2030 on 1990 levels.

This will mean massive investment in renewables like wind farms and strict energy efficiency measures to make business and our homes less wasteful.

The announcement is not only a welcome distraction from Mr Huhne's current personal problems but a political victory after a rift in Cabinet in which some ministers opposed the proposals.

George Osborne, the Chancellor, and his own Lib Dem colleague Vince Cable, the Business Secretary, complained that the plans would damage the manufacturing industry by bringing in green taxes.

But in the end David Cameron intervened to support the ambitious cuts.

There is already a carbon budget in place to cut emissions by 34 per cent by 2020

Speaking in support of his beleaguered minister, the Prime Minister said Number 10 had helped to bring together the warring Cabinet.

'Britain is now leading the world in setting out how we are going to get to a 50 per cent reduction in emissions over the years ahead,' he said.

The Climate Change Act of 2008 sets a target to reduce greenhouse gas emissions in the UK by at least 80 per cent on 1990 levels by 2050, and also requires the Government to set carbon budgets.

The Committee on Climate Change will now monitor the Government's efforts to keep to the latest budget beyond 2020. There is already a carbon budget in place to cut emissions by 34 per cent by 2020.

Analysts said the UK will have to generate 97 per cent of electricity from low carbon sources like nuclear or wind, insulate 3.5 million homes and ensure 60 per cent of new cars run on electricity by 2030 to meet the target. Green taxes will force industry and business to cut carbon.

Terry Scuoler, chief executive of EEF, which represents manufacturing in Britain, warned the businesses will go abroad where rules are not so tough on carbon emissions.

'On its own this is a bad decision for manufacturing so the Government must move quickly to address the competitiveness concerns faced across manufacturing, as well as energy-intensive industries.'

However, Mr Huhne promised a support package for industry to help them move to renewable or low energy power.

He also said there will be a review of the targets in 2014 to ensure the rest of Europe is also introducing measures to cut carbon.

But Andy Atkins, director of Friends of the Earth, said the review was unnecessary.

'The inclusion of a "get-out clause", in case Europe doesn't cut emissions fast enough, creates needless uncertainty that could dent business confidence – and all just to save face for the Chancellor and Business Secretary, who opposed this agreement,' he said.

17 May 2011

INSTITUTE FOR PUBLIC POLICY RESEARCH / THE TELEGRAPH

The Kyoto Protocol

Information from Learning and Teaching Scotland.

The Kyoto Protocol was the world's first international agreement on how to tackle climate change, and an important tool that governments around the world have used since it was made law in 2005. By 2009, 183 countries had signed up to the Protocol and had made a commitment to reduce their carbon dioxide emissions and five other greenhouse gases by an average of 5.2%.

Many countries set their own targets. In the EU this was originally 8% but later increased to 20% by 2020, as governments began to realise that much more had to be done. In the UK and Scotland, Climate Bills more recently committed to reductions of 80%.

Stopping dangerous climate change

The main aim of the Kyoto Protocol was to hold greenhouse gases at a level that will stop dangerous changes to the planet's climate system. All of the industrialised nations that signed and ratified the Protocol would collectively reduce their emissions.

Common problem but different responsibilities

The Kyoto Protocol recognised that we have a common problem but that not all countries have contributed to this problem in the same way. Some countries, including China and India, were exempted from targets because they were not main contributors during the period of industrialisation that is believed to be the cause of climate change.

ALBANIA

Criticism

The United States and Australia originally opted out of Kyoto because of the exemptions granted to China, India and developing countries. They also claimed, along with some economists, that it would cost jobs and damage their countries' economies. However, Australia later signed Kyoto after a change of leadership in 2007, and the US has recently begun working towards its own climate bill.

Many people criticised Kyoto because its mechanisms created a carbon marketplace, where carbon credits could be traded. This allowed richer nations to avoid cutting their emissions and, in some cases, disguise an increase.

Other major criticisms included that the original targets of 5.2% would make little impact on the main cause of climate change – human-induced emissions.

Has the Kyoto Protocol stopped dangerous climate change?

The first phase of Kyoto is due to expire in 2012. It has had its successes and difficulties, and many countries' emissions have actually increased since ratifying the Kyoto Protocol. Many other countries (such as most European countries, the UK and Scotland) have, however, succeeded in reducing their emissions. Most people agree that the Kyoto Protocol has been an important step towards recognising and tackling the problem of climate change. It put climate change on the worldwide agenda for governments.

Most people also agree that not enough has been done to avoid dangerous climate change and that the original Kyoto targets were not strict enough. Since the first Kyoto, climate science has become more alarming and predictions of warming have become more severe. It is now widely held that temperatures are rising and, to prevent the climate from becoming dangerously unstable, emissions will need to be reduced by at least 80% around the world as quickly as possible.

⇨ The above information is reprinted with kind permission from Learning and Teaching Scotland. Visit http://www.ltscotland.org.uk/exploringclimatechange/response/kyotoprotocol.asp for more information on this and other related topics.

© Learning and Teaching Scotland

Carbon offsets

Information from the David Suzuki Foundation.

What is a carbon offset?

A carbon offset is a credit for greenhouse gas reductions achieved by one party that can be purchased and used to compensate (offset) the emissions of another party. Carbon offsets are typically measured in tonnes of CO_2-equivalents (or CO_2e) and are bought and sold through a number of international brokers, online retailers and trading platforms.

For example, wind energy companies often sell carbon offsets. The wind energy company benefits because the carbon offsets it sells make such projects more economically viable. The buyers of the offsets benefit because they can claim that their purchase resulted in new non-polluting energy, which they can use to mitigate their own greenhouse gas emissions. The buyers may also save money as it may be less expensive for them to purchase offsets than to eliminate their own emissions.

Many types of activities can generate carbon offsets. Renewable energy such as the wind farm example above, or installations of solar, small hydro, geothermal and biomass energy can all create carbon offsets by displacing fossil fuels. Other types of offsets available for sale on the market include those resulting from energy-efficiency projects, methane capture from landfills or livestock, destruction of potent greenhouse gases such as halocarbons, and carbon-sequestration projects (through reforestation, or agriculture) that absorb carbon dioxide from the atmosphere.

Why some carbon offsets are better than others

As with any purchase, buyers need to choose their offsets carefully, particularly as the voluntary offset market is largely unregulated.

One issue to consider is the offset project type. For example, although quite popular, offsets from tree-planting projects are problematic for a number of reasons, including their lack of permanence and the fact that these projects do not address our dependence on fossil fuels. Similarly, offset projects involving the destruction of halocarbon gases such as HFC-23 have been subject to numerous criticisms, including the fact that they actually result in a perverse incentive (because of the sheer volume of offsets – and profits—that they generate) for more of the ozone-depleting gas to be created. The price of offsets from these projects is also so low (due to the very high global warming potential of the gas) that they tend to flood the market and squeeze out more sustainable offset projects, like solar and wind.

Another important issue to consider when purchasing offsets is 'additionality'. An offset project is considered additional if it isn't business as usual. Typically, this means that the project wouldn't have happened without the extra funding from the sale of offsets. Additionality is extremely important, as the entire concept of offsetting – i.e. purchasing greenhouse gas reduction credits from a project elsewhere – is based on the premise that those reductions wouldn't have happened otherwise. Only by buying offsets that have met additionality criteria can you be assured that your purchase is resulting in a net benefit for the climate.

Other criteria of high-quality carbon offsets include: validation and verification of the project by reputable third-parties; steps by the project developer to ensure that each offset is only sold once (e.g. by listing the offsets on a public registry); and systems in place to control 'leakage', where the creation of a GHG reduction in one region causes an unintended increase in GHG emissions somewhere else (e.g. protecting a forest in one location could simply shift logging to a forested area in a new location).

The David Suzuki Foundation and the Pembina Institute have prepared a guide, *Purchasing Carbon Offsets*, to help Canadian consumers, businesses and organisations assess the quality of carbon offsets and the vendors that sell them. The following are some questions potential buyers can ask offset vendors:

⇨ What is/are the specific offset project type(s) (e.g. wind farm, methane capture, etc.) in your portfolio and where are the carbon offset projects located?

⇨ Have your carbon offsets been certified to a recognised standard (Gold Standard, CDM, VCS, Climate Action Reserve, Green-e Climate Protocol for Renewable Energy, etc.) to ensure quality? If so, please list the standard(s).

⇨ What steps have you taken to ensure that the carbon offsets you are selling are additional?

⇨ How do you ensure that the greenhouse gas reductions that your carbon offsets represent are quantified accurately?

⇨ Are 100 per cent of your offsets validated and verified by accredited third parties?

DAVID SUZUKI FOUNDATION

⇨ If you are selling offsets that will be created in the future (i.e. forward crediting), what mechanisms (insurance or otherwise) have you put in place to ensure those offsets will actually be delivered?

⇨ What percentage of your portfolio (by tonnes of CO_2e) is made up of offsets from tree-planting or agricultural soils projects? If it is a significant percentage (more than 20 per cent of your portfolio), how do you address permanence risks?

⇨ Do you use a publicly accessible registry to track your offsets? If yes, please list the website. If not, how do you ensure that your offsets are only sold to one buyer? And do you 'retire' offsets that you sell?

⇨ What is your company doing to educate consumers about climate change and the need for government policy to deal with it?

⇨ Are you a member of the International Carbon Reduction and Offset Alliance (ICROA), which has a Code of Best Practice that members must adhere to?

Standards for carbon offsets

Because it can be difficult for offset buyers to get clear answers to each of the above questions, a good way to ensure that your offset purchase is making a positive contribution to the climate is to purchase offsets that meet recognised standards. Just as consumers can feel confident when purchasing food products that meet strict third-party standards for organic agriculture, standards for carbon offsets provide assurance that certain criteria are met when the offset is developed and sold.

A number of standards exist for carbon offsets, including the VCS, Green-e and The Gold Standard. More standards are being announced regularly, and the World Wildlife Fund has published a comparison of the most common offset standards. Each of these standards differs in key ways, with some being more rigorous than others.

The Gold Standard for carbon offsets

The Gold Standard is widely considered to be the highest standard in the world for carbon offsets. It ensures that key environmental criteria have been met by offset projects that carry its label. Significantly, only offsets from energy efficiency and renewable-energy projects qualify for the Gold Standard, as these projects encourage a shift away from fossil-fuel use and carry inherently low environmental risks. Tree-planting projects are explicitly excluded by The Gold Standard.

First, Gold Standard projects must meet very high additionality criteria to ensure that they contribute to the adoption of additional sustainable-energy projects, rather than simply funding existing projects. The Gold Standard also includes social and environmental indicators to ensure the offset project contributes to sustainable development goals in the country where the project is based. Finally, all Gold Standard projects have been independently verified by a third party to ensure integrity.

Currently, The Gold Standard is restricted to offset projects in countries that don't have emission reduction targets under the Kyoto Protocol, primarily developing countries. Supporting offset projects that meet The Gold Standard therefore helps these countries leapfrog developed countries technologically so they don't go down the same fossil-fuel path, which would be disastrous for the climate.

The Gold Standard is supported by more than 70 non-governmental organisations worldwide, including WWF International, Greenpeace International, the Pembina Institute and the David Suzuki Foundation.

⇨ The above information is reprinted with kind permission from the David Suzuki Foundation. Visit www.davidsuzuki.org for more information.

© David Suzuki Foundation

I GOT A GREAT DEAL WITH CARBON CREDITS AFTER I CONVERTED IT TO PEDAL POWER

TECHNOLOGY SURELY IS A WONDROUS THING!

Key agreements in the history of climate change

Timeline of the most important agreements throughout recent history, by Amy Himsworth.

November 1988 – Formation of the Intergovernmental Panel on Climate Change

The Intergovernmental Panel on Climate Change (IPCC) was formed, made up of leading scientists and climate change experts. It held its first meeting in Geneva and aimed to assess scientific knowledge on climate change, analyse its effects and formulate practical solutions.

August 1990 – Publication of IPCC's First Assessment Report

The IPCC's First Assessment Report concluded that man-made emissions were adding to the natural greenhouse gases in the atmosphere. Unless measures were adopted to limit emissions, this addition would lead to an average increase in global temperature.

June 1992 – Signing of the United Nations Framework Convention on Climate Change

At the Rio Earth Summit, the United Nations Framework Convention on Climate Change (UNFCCC) was signed by 154 countries. The main objective was to stabilise greenhouse gases in the atmosphere 'at a level that would prevent dangerous interference with the climate system'.

December 1997 – Signing of the Kyoto Protocol

More than 150 countries signed the Kyoto Protocol, which bound 38 developed countries to reduce greenhouse gas emissions by at least 5% below 1990 levels for the years 2008-2012. Developing nations, including China, had no formal binding targets.

December 2001 – Rules on implementation of the Kyoto Protocol finalised

The final, detailed rules for the implementation of the Kyoto Protocol were agreed. Australia, the US, Japan and Canada forced the European Union to accept major concessions so that a final agreement could be reached.

February 2005 – The Kyoto Protocol becomes law

The Kyoto Protocol became international law.

July 2009 – G8 countries reach agreement

The G8 countries (US, UK, France, Germany, Italy, Canada, Japan and Russia) decided that a limit of two degrees of average global warming should not be surpassed. In order to reach this goal, global greenhouse gas emissions should be cut by at least 50% by 2050.

December 2009 – UN climate summit in Copenhagen

A disappointing result was achieved at the Copenhagen conference when a weak deal was reached between the US, Brazil, India, South Africa and China. The agreement 'recognised' the importance of limiting the global temperature rise to two degrees, but no legally binding commitments were required.

December 2010 – UN climate change conference in Cancún

Elements agreed by the UN in Cancún included: a Green Climate Fund, which would transfer money from the developed to the developing world to tackle the impacts of climate change; the deployment of money and technology for developing countries to build their own sustainable futures.

April 2011 – UN climate change negotiations in Bangkok

UNFCCC Executive Secretary Christiana Figueres called on governments to tackle work settled in the 2010 Cancún talks and address shortfalls in climate action. Detailed discussions were held on how to implement certain aspects, such as the creation of new technology for developing countries.

May 2011

INDEPENDENCE EDUCATIONAL PUBLISHERS

Rich nations failing to keep Copenhagen promise

Research published today shows that developed nations are failing to keep the promise they made last year to provide adequate finance which would help the world's poorest countries adapt to the impacts of climate change.

The paper – published by the International Institute for Environment and Development (IIED) – includes a five-point plan to enable developed nations to fulfil their pledges and build the trust needed to advance the next session of UN climate change negotiations, which begin on 29 November in Cancún, Mexico.

'In last December's climate summit in Copenhagen the developed countries committed to provide developing nations with US$30 billion between 2010 and 2012, with the money balanced between funding for mitigation and adaptation projects,' says Achala Chandani of IIED. 'Our research shows that the developed countries have failed to meet their responsibility to help poorer nations.'

The research shows that funding pledges made since the Copenhagen meeting are far from balanced, with very little earmarked for projects that would enable developing nations to enhance their resilience to climate change impacts on agriculture, infrastructure, health and livelihoods.

> ## Our research shows that the developed countries have failed to meet their responsibiiity to heip poorer nations

'Only US$3 billion has been formally allocated for adaptation,' says Dr Saleemul Huq of IIED. 'There is also a danger that some of this could come in the form of loans which would further indebt already poor nations and force them to pay to fix a problem that the developed nations created.'

The researchers warn that it is also unclear how the money will be disbursed, what type of projects it will support and how the global community will be able to track adherence to pledges and ensure that the funding is truly new and additional to existing aid budgets.

'Currently there is no common framework to oversee, account for and enforce the delivery of the money that rich nations promised to support adaptation to climate change in developing nations,' says Dr J. Timmons Roberts, Director of the Center for Environmental Studies at Brown University and co-director of the AidData project (www.aiddata.org).

'Industrialised nations seem to think they can get away with an "anything goes" approach – where whatever they describe as adaptation funding counts,' adds Roberts. 'The danger is that existing development projects that are not specific responses to the threat of climate change will simply be relabelled as climate adaptation projects.'

The researchers say that to rebuild trust on both sides of the North-South divide, industrialised countries should support an independent registry under the UN Framework Convention on Climate Change and then provide it with detailed and timely data and on all their climate-related projects.

'We have technology now that would allow recipient governments and civil society groups of all types to add their own information about the progress and effectiveness of every adaptation project planned and underway,' adds Roberts. 'By tracking funds all the way from taxpayers in developed nations to each expenditure in the developing countries, this system could create a new era in global cooperation, avoiding many of the pitfalls of past foreign aid.'

David Ciplet, a researcher at Brown University, adds: 'The big promises for adaptation funding made at Copenhagen are not being met. Rather, a fragmented non-system for deciding what counts as adaptation funding is forming, and there is no way to truly measure whether the promises are being met.'

'Adaptation funding is absolutely crucial for the billions of people who face the rising intensity of climate disasters, but making promises is only the first step,' says Ciplet. 'What matters now is that developed countries make good on their promises and provide the funding needed to enable vulnerable countries and communities to increase their resilience to climatic threats such as droughts and floods, rising sea levels and new risks from diseases and crop pests.'

16 November 2010

⇨ The above information is reprinted with kind permission from the International Institute for Environment and Development (IIED). Visit www.iied. org for more information.

© *International Institute for Environment and Development*

Cancún agreement rescues UN credibility but falls short of saving planet

$100-billion 'green climate fund' committed to help poor countries defend themselves against climate change – money likely to come from private sector.

By Suzanne Goldenberg

The modest deal wrangled out by the 200 countries meeting at the Mexican resort of Cancún may have done more to save a dysfunctional UN negotiating process from collapse than protect the planet against climate change, analysts said today.

'The UN climate talks are off the life-support machine,' said Tim Gore of Oxfam. 'The agreement falls short of the emissions cuts that are needed, but it lays out a path to move towards them.'

The agreement produced in the early hours of Saturday reinforces the promise made by rich countries last year to mobilise billions for a green climate fund to help poor countries defend themselves against climate damage.

It was not clear how the funds would be raised. At Copenhagen last year, rich countries agreed to raise $100 billion (£63 billion) a year by 2020 for the fund. However, US officials said at the weekend that most of this would come from the private sector.

Cancún also produced a victory for forest campaigners who were looking to the talks to produce a system of incentives to prevent the destruction of tropical rainforests in countries such as Brazil, Congo and Indonesia.

Under the deal, developing countries will receive aid for not burning or logging forests. Deforestation produces about 15% of the world's carbon emissions.

But with a widening divide between rich and poor countries over the architecture of a global agreement, Patricia Espinosa, the Mexican foreign minister credited with preventing a collapse of the two-week talks, told negotiators the result was 'the best we could achieve at this point in a long process'.

Negotiators, clean-energy business associations and campaign groups warned that Cancún's most significant result was putting off the tough decisions until next year's UN summit in South Africa.

'The outcome wasn't enough to save the planet,' said Alden Meyer of the Union of Concerned Scientists. 'But it did restore the credibility of the United Nations as a forum where progress can be made.'

The Global Wind Energy Council said Cancún was only counted a success because of the extremely low expectations going into the talks. 'None of the fundamental political, legal and architectural issues that still must be resolved in order to establish an effective global climate regime have been solved,' it said.

Michael Levi, a fellow at the Council on Foreign Relations, warned that the failure to resolve difficult issues at Cancún – especially over the future of the Kyoto Protocol – makes the risks even higher next year.

He wrote on his blog: 'The Cancún result punts the dispute to next year's talks. But that solution will not be available again: the current Kyoto commitments expire at the end of 2012, making the next UN conference the last practical opportunity to seal a new set of Kyoto pledges.'

But negotiators did not have many options. After the failure of the Copenhagen summit last year, a breakdown at Cancún would have condemned the 20-year climate negotiations, Connie Hedegaard, the European Union's Climate Commissioner, told reporters on Saturday.

In the run-up to Cancún, negotiators acknowledged there was no prospect of reaching a new treaty. They hoped instead for progress on the 'building blocks' to a deal, such as detailed agreements on climate finance, preventing deforestation, enabling technology transfer and accounting for emissions cuts by emerging economies such as China and India.

However, even those modest ambitions were put in jeopardy when Japan and then Russia announced they would not sign on to a second term of the Kyoto Protocol unless the world's big emitters, China and the US, were also legally bound to action.

Campaign groups such as Greenpeace also blamed the US for taking a hard line at the talks – partly for fear of being accused of giving up too much to China by Republicans at home.

Despite those tensions, however, America and China avoided the mood of confrontation that undermined the talks at Copenhagen last year.

12 December 2010

THE GUARDIAN

⇨ Climate change on a global scale, whether natural or due to human activity, can be initiated by processes that modify either the amount of energy absorbed from the Sun, or the amount of infrared energy emitted to space. (page 1)

⇨ Current understanding of the physics (and increasingly the chemistry and biology) of the climate system is represented in a mathematical form in climate models, which are used to simulate past climate and provide projections of possible future climate change. (page 2)

⇨ The Earth's climate has always changed naturally in the past. But what is happening now is potentially a big change in the Earth's climate, this time caused mainly by human activity. (page 3)

⇨ In the UK, climate change will lead to warmer winters, but temperatures will become uncomfortably hot in summer, and the climate may also be unpredictable and extreme. There's also the risk of rising sea levels and extreme weather like storms and floods. (page 4)

⇨ How our climate will change depends on the future level of carbon dioxide and other gas emissions in the atmosphere. Some impacts are also highly unpredictable in a complex climatic system. (page 9)

⇨ It is estimated that by 2050 there will be 250 million people who will be forced to flee their homes due to drought, desertification, sea-level rise and extreme weather events. Many human populations on islands in the Pacific have already become victims of climate change. (page 10)

⇨ The first decade of this century has been, by far, the warmest decade on the instrumental record. Despite 1998 being the warmest individual year – the last ten years have clearly been the warmest period in the 160-year record of global surface temperature. (page 20)

⇨ A significant 84% of the British public agreed with the statement that the planet is warming, but only 18% believe human activity is mainly responsible; most (58%)

feel that other factors have a part to play. 8% think that human activity, in comparison to other factors, is not responsible at all. (page 21)

⇨ The first wave of criticisms about climate change surfaced in November 2009 in the run-up to the Copenhagen climate negotiations, when about a thousand private emails were stolen from a server at UEA and released online. (page 22)

⇨ One-third (33%) of the public now agrees with the statement 'it is not yet clear whether climate change is happening or not – scientists are divided on this issue', compared to only 25% in 2007. (page 25)

⇨ Mitigation needs international agreement and national enforcement; emissions from one country threaten everybody, and cutting them will take time. Adaptation, on the other hand, is immediate and local. (page 28)

⇨ The way that humanity reacts to climate change may do more damage to many areas of the planet than climate change itself unless we plan properly, an important new study published by Conservation International's Will Turner and a group of other leading scientists has concluded. (page 31)

⇨ Chris Huhne, the Energy and Climate Change Minister, said the UK would drastically speed up green measures from 2020, ultimately cutting carbon by 60 per cent by 2030 on 1990 levels. This will mean massive investment in renewables like wind farms and strict energy efficiency measures to make business and our homes less wasteful. (page 33)

⇨ The Kyoto Protocol was the world's first international agreement on how to tackle climate change, and an important tool that governments around the world have used since it was made law in 2005. (page 34)

⇨ A carbon offset is a credit for greenhouse gas reductions achieved by one party that can be purchased and used to compensate (offset) the emissions of another party. (page 35)

Adaptation

In relation to climate change, adaptation aims to respond to the effects of global warming by adapting to altered environments. This includes adapting to changed food production methods, agriculture and sea levels.

Carbon footprint

A carbon footprint is a measure of an individual's effect on the environment, taking into account all greenhouse gases that have been emitted for heating, lighting, transport, etc. throughout that individual's average day.

Carbon offsets

Carbon offsets are a reduction in greenhouse gas emissions made in order to compensate for greenhouse gas production somewhere else. Offsets can be purchased in order to comply with caps, such as the Kyoto Protocol. For example, rich industrialised countries may purchase carbon offsets from a developing country in order to satisfy environmental legislation.

Climate change

Climate change describes a global change in the balance of energy absorbed and emitted into the atmosphere. This imbalance can be triggered by natural or human processes. It can cause either regional or global changes in weather averages and frequency of severe climatic events.

Climate change refugees

Also known as 'environmental migrants', climate change refugees are people who have been forced to flee their home region following severe changes in their local environment as a result of global warming.

Climate models

Scientific models which are designed to replicate the Earth's climate. Scientists are able to hypothetically test the effects of global warming by simulating changes to the Earth's atmosphere.

CO_2 emissions

Carbon dioxide gas released into the atmosphere. CO_2 is released when fossil fuels are burnt. An increase in CO_2 emissions due to human activity is arguably the main cause of global warming.

Global warming

This refers to a rise in global average temperatures, caused by higher levels of greenhouse gases entering the atmosphere which then absorb and trap radiative forces. Global warming is affecting the Earth in a number of ways, including melting the polar ice caps, which in turn is leading to rising sea levels.

Greenhouse gases (GHG)

A greenhouse gas is a type of gas that can absorb and emit longwave radiation within the atmosphere: for example, carbon dioxide, methane and nitrous oxide. Human activity is increasing the level of greenhouse gases in the atmosphere, causing the warming of the Earth. This is known as the greenhouse effect.

IPCC

An abbreviation for Intergovernmental Panel on Climate Change, the leading scientific body which assesses and reviews global climate change. It was founded by the United Nations Environment Programme and the World Meteorological Organization and currently has 194 member countries from around the world.

Kyoto Protocol

An international treaty setting binding targets for 37 developed countries to reduce their greenhouse gas emissions by at least five per cent below 1990 levels for the years 2008-2012. It was made international law in 2005. It was the world's first international agreement on tackling climate change.

Mitigation

In relation to climate change, mitigation refers to the act of reducing, or limiting, the level of greenhouse gas emissions in order to slow the rate of global warming. Emissions targets, government campaigns and the development of 'greener' energy sources are examples of how mitigation can be used to reduce climate change.

REDD

An abbreviation for the United Nations programme Reducing Emissions from Deforestation and Forest Degradation. This collaborative initiative aims to assist developing countries in combatting deforestation, illegal logging and fires in order to limit climate change.

ACKNOWLEDGEMENTS

The publisher is grateful for permission to reproduce the following material.

While every care has been taken to trace and acknowledge copyright, the publisher tenders its apology for any accidental infringement or where copyright has proved untraceable. The publisher would be pleased to come to a suitable arrangement in any such case with the rightful owner.

Chapter One: The Climate Crisis

Climate and climate change: some background science, © The Royal Society, *Myths about climate change*, © Crown copyright is reproduced with the permission of Her Majesty's Stationery Office, *The status of climate change science today*, © UNFCC, *Abrupt climate change*, © David Suzuki Foundation, *What about climate change in the future?*, © National Trust, *The social and economic impacts of climate change*, © Earthwatch, *West Atlantic Ice Sheet 'could be more stable than thought'*, © University of Exeter, *Top ten global weather/climate events of 2010*, © National Climatic Data Center (NOAA), *Climate change: the forest connection*, © SinksWatch, *Greenland ice sheet faces 'tipping point in ten years'*, © Guardian News and Media Limited, *The complicated truth about sea-level rise*, © CICERO.

Chapter Two: The Climate Debate

Unscientific hype about the flooding risks from climate change will cost us all dear, © Telegraph Media Group Limited, *Ten facts on climate science*, © Crown copyright is reproduced with the permission of Her Majesty's Stationery Office, *Climate change blame?*, © YouGov, *Are we still sure about climate change?*, © Scientists for Global Responsibility, *Climate change scepticism*, © YouGov, *Can we really measure the climate?*, © The Scientific Alliance.

Chapter Three: Policies and Solutions

Adapting to the greenhouse, © Panos London, *We have to adapt culturally to climate change*, © Dr Jenny Pickerill, *Saving the world from the human response to climate change*, © Conservation International, *Technology, not targets, will save the planet*, © Institute for Public Policy Research, *Climate change victory for Huhne*, © Telegraph Media Group Limited, *The Kyoto Protocol*, © Learning and Teaching Scotland, *Carbon offsets*, © David Suzuki Foundation, *Key agreements in the history of climate change*, © Independence Educational Publishers, *Rich nations failing to keep Copenhagen promise*, © International Institute for Environment and Development, *Cancún agreement rescues UN credibility but falls short of saving the planet*, © Guardian News and Media Limited.

Illustrations

Pages 3, 17, 20, 30: Angelo Madrid; pages 8, 34: Bev Aisbett; pages 9, 18, 23: Simon Kneebone; pages 11, 21, 26, 36: Don Hatcher.

Cover photography

Left: © Piotr Błoch. Centre: © BSK. Right: © Ansgar Walk.

Additional acknowledgements

Editorial by Carolyn Kirby on behalf of Independence.

And with thanks to the Independence team: Mary Chapman, Sandra Dennis and Jan Sunderland.

Lisa Firth
Cambridge
September, 2011

south essex college
FURTHER & HIGHER EDUCATION
SOUTHEND CAMPUS

The following tasks aim to help you think through the issues surrounding the climate change debate and provide a better understanding of the topic.

1 Climate change is a phrase used widely in modern society, particularly by the mass media and politicians, but what exactly is it? Write a brief summary of what climate change is and the processes which cause it. You can use the article *Climate and climate change: some background science* on page 1 to help you.

2 In small groups, role play a radio talk show on the topic of climate change. One student will play the radio show host, another a climate change expert who aims to dispel some common myths, and a third student will be a climate change sceptic arguing that human actions do not cause global warming. Other students can play listeners phoning in with questions. The host should aim to stimulate a lively debate on the topic, giving equal time to all arguments.

3 Design and present a ten-minute PowerPoint presentation on the predicted future impacts of climate change. Explain what is expected to happen to global temperatures and sea levels. You can use *What about climate change in the future?* on page 9 to help you.

4 Watch Al Gore's 2006 documentary film 'An Inconvenient Truth'. Do you find his arguments convincing? Do you think this film is an effective way of conveying the climate change message? Write a review of his presentation.

5 There is already evidence that climate change is happening; however, there is some debate over the extent to which human activity is causing or exacerbating this. Using the information in Chapter Two as well as your own research, summarise the arguments for and against the theory that human activity is the main reason for climate change. Create two lists, one supporting and one opposing the theory. Which list is longer? Which do you agree with?

6 What is your carbon footprint? Visit www.carboncalculator.direct.gov.uk to find out your carbon footprint. How could you reduce this? Compare your findings with others in your class. Is there much variation throughout your class or do you all have similar footprints?

7 Study the graphs on page 24. Write a summary of the conclusions which can be drawn from these graphs. What do they show? Try to find three points to discuss then compare your findings with others in your class.

8 Read *Can we really measure the climate?* on page 26. In pairs, discuss how valid you think the author's argument is. Do you agree that we should dismiss climate predictions because of flawed scientific methods? Or do you think averages are the best measure we have, even if they are not perfect? Try to assess the debate evenly, weighing up the pros and cons of each argument. What is your conclusion?

9 Climate change solutions fall into two categories: adaptation and mitigation. Write an informative article explaining the differences between mitigation and adaptation. What are the advantages and disadvantages of each?

10 Design a wall poster giving information on the Kyoto Protocol. Use a mind map or a spider diagram design to display key facts and figures about the Protocol. Include answers to questions such as: what is the Kyoto Protocol? When was did it become law? Who does it affect? Do you think it is successful in reducing greenhouse gas levels?

11 Write an informative leaflet explaining what carbon offsets are and how they work. Include information on who might purchase carbon offsets, and how. You should make it readable and attractive, suitable for a teenage audience. You can add illustrations or diagrams to your leaflet to make it more accessible.

12 In his article *Technology, not targets, will save the planet* on page 32, Andrew Pendleton argues that 'Innovation, not emissions targets, should be the prime focus of international efforts.' Discuss this statement in groups. Do you agree with Pendleton? Do you think legislation can address climate change or should governments focus on developing 'greener technology'? Write a summary of your conclusions.